AF842252

Objective Agronomy

For SAUs Entrance, JRF, SRF, ARS, NET, Ph.D, Civil Services and other Competitive Examination

THE AUTHOR

Dr. B.B. Singh (M.Sc., Ph.D., PGDEM, FISEP, FLWS) obtained M.Sc. in 1992 from Dr. R.M.L. Avadh University, Faizabad and Ph.D. in 1996 from D.D.U. Gorakhpur University, Gorakhpur. Dr. Singh has worked in the capacity of S.R.F. (ICAR), Research Associate (UGC) and Faculty at N.D. University of Agriculture and Technology, Kumarganj, Faizabad (U.P.). He was actively engaged in teaching and research at B.Sc. (Ag.), M.Sc. (Ag.) and Ph.D. levels.

Dr. Singh is the first recipient of ***'Young Scientist Award'*** by International Society for Environmental Protection, India in the year 1997, besides this he has been the recipient of many honors, awards and distinctions from various scientific bodies. He has published several research papers in journals of repute and also authored more than 24 books. Dr. Singh is honorary Referee for evaluating research papers in many scientific journals. He has also been honored with Fellowships *viz.* Fellow International Society for Environmental Protection (FISEP) and Fellow Living World Society (FLWS). He has participated/delivered lectures at various National and International conferences/symposiums etc.

Dr. Singh has delivered several Radio talks on different aspects of Agriculture and at presently he is actively involved in writing and editing of books.

Objective Agronomy

For SAUs Entrance, JRF, SRF, ARS, NET, Ph.D, Civil Services and other Competitive Examination

–Author–

Dr. B.B. Singh

M.Sc., Ph.D., PGDEM, FISEP, FLWS

Former Faculty

N.D. University of Agriculture and Technology, Kumarganj

Faizabad (UP)

2016

Daya Publishing House®

A Division of

Astral International (P) Ltd.

New Delhi - 110 002

© 2016 PUBLISHER

Publisher's note:
Every possible effort has been made to ensure that the information contained in this book is accurate at the time of going to press, and the publisher and author cannot accept responsibility for any errors or omissions, however caused. No responsibility for loss or damage occasioned to any person acting, or refraining from action, as a result of the material in this publication can be accepted by the editor, the publisher or the author. The Publisher is not associated with any product or vendor mentioned in the book. The contents of this work are intended to further general scientific research, understanding and discussion only. Readers should consult with a specialist where appropriate.
Every effort has been made to trace the owners of copyright material used in this book, if any. The author and the publisher will be grateful for any omission brought to their notice for acknowledgement in the future editions of the book.
All Rights reserved under International Copyright Conventions. No part of this publication may be reproduced, stored in a retrieval system, or transmitted in any form or by any means, electronic, mechanical, photocopying, recording or otherwise without the prior written consent of the publisher and the copyright owner.

Cataloging in Publication Data—DK
Courtesy: D.K. Agencies (P) Ltd. <docinfo@dkagencies.com>

Singh, B. B., 1948- author.
Objective agronomy : for SAUs entrance, JRF, SRF, ARS, NET, Ph.D, civil services and other competitive examination / author, Dr. B.B. Singh.
pages cm
ISBN 9789351309192 (International Edition)

1. Agronomy—Examinations, questions, etc. I. Title.

DDC 630.76 23

Published by : **Daya Publishing House®**
A Division of
Astral International Pvt. Ltd.
ISO 9001:2008 Certi ed Company
4760-61/23, Ansari Road, Darya Ganj
New Delhi-110 002
Ph. 011-43549197, 23278134
E-mail: info@astralint.com
Website: www.astralint.com

Laser Typesetting : **SSMG Computer Graphics, Burari-110084**

Printed at : **Thomson Press India Limited**

भारतीय कृषि अनुसंधान परिषद
कृषि अनुसंधान भवन-II, पूसा, नई दिल्ली 110 012
INDIAN COUNCIL OF AGRICULTURAL RESEARCH
KRISHI ANUSANDHAN BHAVAN-II, PUSA, NEW DELHI 110 012

डॉ. अरविन्द कुमार
उप महानिदेशक (शिक्षा)
Dr. Arvind Kumar
Deputy Director General (Education)

Phone : 011-25841760 (O)
: 011-25843644 (R)
Fax : 011-25843932
Email : ddgedn@icar.org.in
Website : www.icar.org.in

Foreword-I

Feeding the ever growing population is a cause of concern considering the decline in the natural resources including soil and water. Over the years the net cultivated area is static and the input use efficiency is decreasing. The traditional practices of cultivation are no longer viable. Improving per unit productivity through sound management practices and improved deeds is the present day need to match the growing demand for food, feed and fibre.

The crop management and post-harvest technology have gradually changed. It is essential toww have a sound knowledge of different aspects of agronomy, which deals primarily with crop production.

The present book entitled *"Objective Agronomy"* is a compilation of work and embodies a fairly comprehensive treatment of objective type questions on agronomy. It contains 18 chapters covering all related disciplines. These chapters includes: Introduction, Agricultural Meteorology, Soil Properties and Tillage, Soil and Nutrients Management, Irrigation Water Management, Soil and Water Conservation, Manures and Fertilizers, Cropping Systems, Weed Management, Dryland Management, Cereal Crop Production, Pulse Crop Production, Oil Seed Crop Production, Tuber Crop Production, Sugar Crop Production, Jute and Allied Fibers, Forage Crop Production and Plantation Crop Production.

I am sure that the present book will serve as a ready reference for the aspirants of various competitive examination conducted by public and private sector.

I congratulate Dr. Singh for his sincere efforts in bringing out this comprehensive publication.

(Arvind Kumar)

DDG (Edn.)

भा.कृ.अ.प. – भारतीय कृषि अनुसंधान संस्थान
नई दिल्ली-110012 (भारत)
ICAR -INDIAN COUNCIL OF AGRICULTURAL RESEARCH INSTITUTE
(A UNIVERSITY UNDER SECTION 3 OF UGC ACT, 1956)
NEW DELHI - 110012 (INDIA)

डॉ. रविन्द्र कौर
निदेशक (कार्यकारी)
Dr. (Mrs.) Ravinder Kaur
Director (Acting)

Phones : 011-2584 2367
: 2584 3375 (Off.)
Fax : 011-2584 6420
Email : director@iari.res.in
Personal : rk132.iari@gmail.com
Website : www.iari.res.in

Foreword-II

Agronomy is a dynamic discipline i.e., a synthesis of several disciplines like soil science, agricultural chemistry, crop physiology, plant ecology, biochemistry and economics. Soil physical, chemical and biological properties have to be understood thoroughly for effective modification of soil environment. Similarly, it is necessary to understand the physiology of crops to meet there requirements. Advances in economic analysis helped in production of crops economically along with the advancement of knowledge and better understanding of plant and new practices are developed for high productivity. Availability of chemical fertilizers has necessitated the generation of knowledge on the method, quantity and time of application of fertilizers. Similarly availability of herbicides for the control of weeds has led to development of knowledge about selectivity, time and method of application of herbicides. To overcome the problems, different management practices are developed.

New technology has to be developed to overcome the effect of moisture stress under dryland condition. As new varieties of crops with high yield potential becomes available, package of practices have to be developed to exploit their full potential. Restoration of soil fertility, preparation of good seed bed, use of proper seed rates current dates of sowing for each improved variety, proper conservation and management of soil moisture and proper control of weeds are agronomic practices to make our limited land and water resources more productive.

The book entitled *"**Objective Agronomy**"* by Dr. B.B. Singh is well written and it includes the objective type questions, fill in the blanks and true/false with key towards preparation for competitive examination *viz.* ICAR-JRF/SRF, NET, ARS, SAU's and others related examinations in India. Hence, it is felt that this book can serve as single platform for preparation of competitive examination in the field Agronomy discipline. The questions framed in the book are so clearly asked that one can clearly understand.

It is believed that the readers will get good knowledge and the diminutive doubts will also be cleared after finishing this book. I congratulate Dr. Singh for bringing out this book, which is well written for the students of competitive examination.

Ravinder Kaur

Preface

Agronomy is a science that helps to feed the world. We can call the Agronomy as backbone of all agricultural sciences, because the management of soil and water, with a view to achieve the production potential of high yielding varieties in green revolution is an agronomic domain. At present, there are many Objective Agronomy books available but this is the book from which one can have at least an overview of all aspects of Agronomy.

Objective Agronomy for competitive exams in agricultural discipline, contains 18 chapters covering all related disciplines. These chapters include : Introduction, Agricultural Meteorology, Soil Properties and Tillage, Soil and Nutrients Management, Irrigation Water Management, Soil and Water Conservation, Manures and Fertilizers, Cropping Systems, Weed Management, Dryland Management, Cereal Crop Production, Pulse Crop Production, Oil Seed Crop Production, Tuber Crop Production, Sugar Crop Production, Jute and Allied Fibres, Forage Crop Production and Plantation Crop Production.

This book is primarily intended to serve as a ready reference for those appearing in competitive examination like undergraduate, post-graduate, SAUs Entrance, Ph.D., Junior Research Fellowship (JRF), Senior Research Fellowship (SRF), Agricultural Research Service (ARS), National Eligibility Test (NET), SET, Civil Services, Agricultural Officers, Bank P.O. (Agriculture) and Allied Agricultural Examinations.

I believe that this book will be also useful for fellow teachers, researchers and development officers for references and easy answering of many complicated questions. Every possible effort has been made to accommodate previous years questions from various competitive examinations on memory basis.

I am grateful to all those persons as well as various books, manuals, periodicals, magazines, newsletters, journals etc. that helped in the preparation of this book.

In spite of the best efforts, it is possible that some errors may have occurred into the compilation and editing of the book. Further queries, suggestions and criticisms for improvement of the book are always welcome and shall be thankfully acknowledged. Last but not the least, it is a pleasure for us to extend our sincere thanks to Mr. Anil Mittal, Astral International Pvt. Ltd. and his team members for his keen interest shown in preparation and publishing of this book so efficiently and promptly.

Dr. B.B. Singh

Contents

1

Agronomy : An Introduction

1. **First agronomic is research was conducted at __________.**

   (a) India (b) USA

   (c) China (d) Assells

2. **Agronomy is a __________.**

   (a) Basic Science (b) Information Science

   (c) Medical Science (d) Applied Science

3. **The word 'Agriculture' is derived from __________.**

   (a) Latin language (b) Greek language

   (c) English language (d) None of these

4. **The word 'Agronomy' is derived from _________.**

   (a) French language (b) Latin language

   (c) Greek language (d) English language

5. **Sustainable agriculture can be achieved by _________.**

   (a) More technological support

   (b) More subsidies

   (c) Better prices of the produce

   (d) More resources

6. **Separate Central Agriculture Department was started in the year**

   (a) 1907 (b) 1895

   (c) 1881 (d) 1878

7. **First synthetic fertilizer industry was established by**

   (a) Muller (b) Owen

   (c) Lawes (d) Sutton

8. **Iron Share was invented by**
   (a) Robert Ransome (b) Liebig
   (c) Paul Muller (d) Arthur Young

9. **Round revolution is related to**
   (a) Egg (b) Milk
   (c) Poultry (d) Potato

10. **Share of Agriculture in total GDP of India is about**
   (a) 20% (b) 25%
   (c) 35% (d) 45%

11. **Agriculture which encompassing of**
   (a) Crop production and livestock farming
   (b) Fisheries
   (c) Forestry
   (d) All the above

12. **________ is a branch of agricultural science which deals with principles and practices of soil, water and crop management.**
   (a) Meteorology (b) Soil Science
   (c) Agronomy (d) Agricultural Microbiology

13. **Environment of crop constitutes**
   (a) Soil (b) Aerial environment
   (c) Both (a) and (b) (d) None of these

14. **Aerial environment**
   (a) Can be changed easily (b) Cannot be changed easily
   (c) Can be changed partially (d) None of these

15. **Correct order of domestication of animal is**
   (a) Sheep, goat, dog (b) Dog, goat, sheep
   (c) Dog, sheep, goat (d) Goat, sheep, dog

16. **Bloodless revolution pertains to which one of the following**
   (a) White revolution (b) Agriculture reforms
   (c) Land donations (d) Green revolution

17. **The basic principles of Agronomy is/are ________.**
   (a) Management of soil and climate
   (b) Appropriate inter-cropping activities management for crop
   (c) To select appropriate materials for seed and sowing
   (d) All of these

18. The highest nitrogen content is found in ________.

(a) Farmyard manure (b) Vermi compost
(c) Poultry manure (d) Biogas slurry

19. Harmful substance in urea is __________.

(a) Cyanide (b) Amide
(c) Uric acid (d) Biuret

20. First cultivated crops in the world are

(a) Rice, Pea (b) Wheat, Mustard
(c) Wheat, Barley (d) None of these

21. Plough was invented during which period?

(a) 2900 BC (b) 3900 BC
(c) 2000 BC (d) 4900 BC

22. Earliest rice cultivation was noticed in which one of the following

(a) Indus valley (b) Indo-Gangetic plains
(c) Deccan plateau (d) None of these

23. Theory of 'Humus' was explained in the year

(a) 1809 (b) 1819
(c) 1920 (d) 1900

24. In the world Department of Agriculture was first established in the year

(a) 1862 (b) 1882
(c) 1850 (d) 1890

25. Agriculture Department in India was established in the year

(a) 1860 (b) 1880
(c) 1890 (d) 1871

26. In India Department of Agriculture was established by

(a) Lord Cornnwallis (b) Lord Mayo
(c) Lord Curzon (d) None of these

27. State Seed Farm Corporation was established in the year

(a) 1969 (b) 1975
(c) 1999 (d) 2000

28. Per capita availability of food grains in India is

(a) 400 g (b) 800 g
(c) 500 g (d) 200 g

29. Per capita availability of pulses in India is

(a) 36 g (b) 45 g
(c) 55 g (d) 65 g

30. Area under shifting cultivation in India is around

(a) 5 mha (b) 6 mha
(c) 7 mha (d) 8 mha

31. Per capita energy required for man is

(a) 2500 k cal (b) 2200 k cal
(c) 3500 k cal (d) 3000 k cal

32. Percentage of India's geographical area to the world is

(a) 6 (b) 2
(c) 4 (d) 5

33. The capacity of soil to produce the yield is called

(a) Soil index (b) Soil productivity
(c) Soil capacity (d) Soil fertility

34. The study of soil in relation to plant growth is called __________.

(a) Pedology (b) Histology
(c) Edaphology (d) Anthrology

35. Rural Electrification Corporation was set up in India in the year

(a) 1956 (b) 1975
(c) 1969 (d) 1978

36. Which among the state is having the largest number of electric pumps?

(a) Bihar (b) UP
(c) Uttarakhand (d) Tamil Nadu

37. National Seed Corporation was set up in India in the year

(a) 1960 (b) 1963
(c) 1966 (d) 1969

38. National Seed Project was established in the year

(a) 1960 (b) 1988
(c) 1998 (d) 1976

39. The separate branch of botany was originated from agriculture in __________.

(a) 1910 (b) 1900
(c) 1950 (d) 1952

40. Principles of conservation agriculture is/are ________.

(a) Precision land levelling
(b) Minimal disturbances of the soil
(c) Permanent soil cover with organic materials
(d) All of these

41. Nitrification bacteria was isolated by

(a) Warrington (b) Winogradsky
(c) Beijerinck (d) Liebig

42. First Agronomic Research was started by

(a) Liebig (b) J.B. Bossinggault
(c) B.P. Pal (d) Mather

43. American Society of Agronomy is established in year ___________.

(a) 1908 (b) 1947
(c) 1975 (d) None of these

44. The production system that sustain the health of soils, ecosystems and peoples is called ___________.

(a) Organic agriculture (b) Industrial agriculture
(c) Conservation agriculture (d) Sustainable agriculture

45. During stress which of the following phytohormone is accumulated ?

(a) Auxin (b) ABA
(c) Ethylene (d) OA

46. The plants grow in hydric conditions are called

(a) Halophytes (b) Tenagophytes
(c) Hydrophytes (d) None of these

47. The status of soil water is also represented in term of

(a) Field capacity (b) Hygroscopic coefficient
(c) Wilting coefficient (d) Capillary potential

48. The catchment area where ground water reaches the surface and run-off is termed as

(a) Run-off (b) Springs
(c) Aquifer (d) None of these

49. Crop highly manipulated ecosystem is fed to domesticated animal is called _______.

(a) Animal agriculture (b) Shifting cultivation
(c) Terrestrial agriculture (d) Livestock management

50. Earth is an open system for :

(a) Gases (b) Water
(c) Minerals (d) Energy

51. Father of Modern Agronomy is ________.

(a) K.C. Mehta (b) Peter Dearesenzi
(c) E.J. Butler (d) Gregor John Mendel

52. Agriculture is a

(a) Physical process (b) Chemical process
(c) Biological process (d) All of these

53. In the following which is not a soil chemical environment ?

(a) Soil nutrient (b) Soil pH
(c) Soil Electrical conductivity (d) Soil water

54. Which among the following is not amenable to modification ?

(a) Soil environment (b) Aerial environment
(c) Both (a) and (b) (d) None of these

55. In which BC cultivation of Wheat and Barley begins ?

(a) 8700 BC (b) 7500 BC
(c) 7700 BC (d) 5800 BC

56. In which BC cultivation of Maize begins ?

(a) 8700 BC (b) 4400 BC
(c) 7700 BC (d) 5800 BC

57. Primary resource for agriculture is

(a) Air (b) Tractor
(c) Fertilizer (d) Pesticide

58. Primary resources for agriculture are

(a) Land, Air and Water
(b) Seed, Fertilizer and Pesticide
(c) Seed, Air and Water
(d) Land, Fertilizer and Water

59. In the following which is not a component of aerial environment for agriculture ?

(a) Wind (b) Solar radiation
(c) Relative humidity (d) None of these

60. Component of aerial environment for agriculture is

(a) Rainfall (b) Temperature
(c) Relative humidity (d) All of these

61. ***Kharif*, *Rabi* and *Zaid* words belong to :**

(a) German (b) Spanish
(c) Arabic (d) French

62. Rosthamsted Agricultural Experiment Station was established by :

(a) E.J. Butler (b) John Bennet Lawes
(c) Justus von Liebig (d) Peter Dearesenzi

63. Natural farming was advocated by :

(a) E.J. Butler (b) Fakuoka
(c) M.S. Swaminathan (d) Norman E. Borlaug

64. Imperial Agricultural Research Institute was established in

(a) 1905 (b) 1900
(c) 1940 (d) 1930

65. Imperial Agricultural Research Institute was located at

(a) Pusa (b) Kolkata
(c) Lucknow (d) Nagpur

66. Of the total volume of inland water ____ is in the form of ground water

(a) 10% (b) 78%
(c) 80% (d) 22%

67. Of the total volume of inland water ____ is in the form of fresh water

(a) 1.0% (b) 75%
(c) 2.5% (d) 23%

68. Total irrigated area world wide is about ________ .

(a) 320 Mha (b) 210 Mha
(c) 300 Mha (d) 255 Mha

69. The total geographical area of India is about ________ .

(a) 1400 Mha (b) 2000 Mha
(c) 424 Mha (d) 328 Mha

70. The total area subjected to soil erosion and degradation in India is about ________ .

(a) 150 Mha (b) 188 Mha
(c) 134 Mha (d) 330 Mha

71. The total food loss due to weeds is estimated to be _______ globally .

(a) 80 Mt (b) 140 Mt
(c) 100 Mt (d) 287 Mt

72. In India the weeds accounts for ________ of the total food loss.

(a) 65% (b) 100%
(c) 40% (d) 45%

73. In India the insects accounts for ________ of the total food loss.

(a) 85% (b) 30%
(c) 55% (d) 40%

74. In India the diseases accounts for _________ of the total food loss.

(a) 20% (b) 40%
(c) 10% (d) 60%

75. The weeds accounts for _________ % reduction in yield of sugarbeet.

(a) 70% (b) 40%
(c) 50% (d) 30%

76. Central Arid Zone Research Institute is situated at

(a) Jodhpur (b) Kanpur
(c) Jhansi (d) Nagpur

77. Central Institute for Cotton Research is situated at

(a) Barrackpore (b) Nagpur
(c) Shimla (d) New Delhi

78. Central Institute for Sub-tropical Horticulture is situated at

(a) Bhopal (b) Nagpur
(c) Lucknow (d) Shimla

79. Directorate of Wheat Research is situated at

(a) Kanpur (b) Nagpur
(c) Hyderabad (d) Karnal

80. Directorate of Water Management Research is situated at

(a) Pune (b) Pusa, New Delhi
(c) Jhansi (d) Ranchi

81. Indian Agricultural Research Institute (Deemed University) is situated at

(a) Rajahmundry (b) New Delhi
(c) Nagpur (d) Karnal

82. Indian Grassland and Fodder Research Institute is situated at

(a) Kasargod (b) Mumbai
(c) Jhansi (d) Karnal

83. Indian Institute of Sugarcane Research is situated at

(a) Nagpur (b) Mumbai
(c) Hyderabad (d) Lucknow

84. Indian Institute of Soil Science is situated at

(a) Bhopal (b) Bikaner
(c) Meghalaya (d) New Delhi

85. National Centre for Mushroom Research and Training is situated at

(a) New Delhi (b) Lucknow
(c) Junagadh (d) Solan

86. National Research Centre for Cashew is situated at

(a) Patna, Bihar (b) Nagpur, Maharashtra
(c) Calcutta, W.B. (d) Puttar, Karnataka

87. National Research Centre for Groundnut is situated at

(a) Varanasi (b) Bhopal
(c) Junagadh (d) Kanpur

88. National Research Centre on Soybean is situated at

(a) Faridabad (b) Indore
(c) New Delhi (d) Varanasi

89. National Research Centre on Citrus is situated at

(a) Nagpur (b) New Delhi
(c) Pune (d) Kanpur

90. Central Research Institute for Dry lands Agriculture is situated at

(a) Hyderabad (b) Kanpur
(c) Varanasi (d) Karnal

91. Central Rice Research Institute is situated at

(a) Nagpur (b) Cuttack
(c) Dehradun (d) Kasargod

92. Central Soil and Water Conservation Research and Training Institute is situated at

(a) Hyderabad (b) New Delhi
(c) Dehradun (d) Karnal

93. Central Soil Salinity Research Institue is situated at

(a) Solan (b) Cuttack
(c) Shimla (d) Karnal

94. Central Tobacco Research Institute is situated at

(a) Rajahmundry (b) Jhansi
(c) Patna (d) Junagadh

95. Central Institute for Research on Cotton Technology is situated at

(a) Rajahmundry (b) Mumbai
(c) Hyderabad (d) Rahuri

96. In total irrigated area in the world, India share is

(a) 20% (b) 40%
(c) 10% (d) 30%

97. In the world Wheat production, India ranks ______

(a) First (b) Second
(c) Third (d) Fourth

98. In the world Rice production, India ranks ______

(a) First (b) Second
(c) Third (d) Fourth

99. India ranks _______ in cotton production in the world

(a) First (b) Second
(c) Third (d) Fourth

100. In the world Sugarcane production, India ranks ______

(a) First (b) Second
(c) Third (d) Fourth

101. NSC was established in the year

(a) 1970 (b) 1963
(c) 1961 (d) 1985

102. In India the apex agency for coordinating agricultural research and education is

(a) IARI (b) ICAR
(c) SAUs (d) All of these

103. The cultivable per capita land availability world wide is about ____ ha.

(a) 0.3 (b) 0.8
(c) 1.5 (d) 0.7

104. The cultivable per capita land availability in India is about ______ ha.

(a) 0.04 (b) 3.0
(c) 0.15 (d) 1.5

105. The total global volume of water is about _____ million cubic kilometers

(a) 1400 (b) 1000
(c) 355 (d) 410

106. The total volume of water in the sea is about _______.

(a) 4% (b) 87%
(c) 88% (d) 97%

107. The total volume of water in inland is about _______ .

(a) 3% (b) 87%
(c) 97% (d) 33%

108. Of the total volume of inland water ____ is in the form of glaciers

(a) 20% (b) 77%
(c) 88% (d) 33%

109. ________ of total area subjected to soil erosion and degradation in India

(a) 50% (b) 88%

(c) 28% (d) 54%

110. The total irrigated area of India is about _______

(a) 50 Mha (b) 30 Mha

(c) 40 Mha (d) 330 Mha

111. In India dryland contribute about _________ % of the cultivated area

(a) 72 (b) 38

(c) 82 (d) 48

112. In India dryland contribute about ______ % of the total food production

(a) 72 (b) 38

(c) 43 (d) 28

113. In total irrigated area in the world, India ranks

(a) Fourth (b) Second

(c) Third (d) First

114. In the world Pulses production, India ranks

(a) First (b) Second

(c) Third (d) Fourth

115. In the world Groundnut production, India ranks

(a) First (b) Second

(c) Third (d) Fourth

116. India ranks _______ in Tea production in the world

(a) First (b) Second

(c) Third (d) Fourth

117. India ranks _______ in Milk production in the world

(a) First (b) Second

(c) Third (d) Fourth

118. Fertilizer consumption in India is about _______ .

(a) 75 kg/ha (b) 140 kg/ha

(c) 155 kg/ha (d) 220 kg/ha

119. Maximum fertilizer consumption in the world is in

(a) Netherlands (b) India

(c) China (d) Pakistan

120. Maximum reduction in yield due to weeds has been observed in

(a) Sugarbeet (b) Pea

(c) Mustard (d) Gram

121. Central Potato Research Institute is situated at

(a) Nainital (b) Nagpur
(c) Lucknow (d) Shimla

122. Central Research Institute for Jute and Allied Fibres is situated at

(a) Barrackpore (b) Nagpur
(c) Kanpur (d) Mumbai

123. Central Agricultural Research Institute is situated at

(a) Hyderabad (b) Port Blair
(c) Nagpur (d) Kasargod

124. Central Institute of Agricultural Engineering is situated at

(a) Barrackpore (b) Port Blair
(c) Bhopal (d) Kasargod

125. Central Plantation Crops Research Institute is situated at

(a) Cuttack (b) Karnal
(c) Bhopal (d) Kasargod

126. Directorate of Oilseeds Research is situated at

(a) New Delhi (b) Mumbai
(c) Hyderabad (d) Karnal

127. Directorate of Rice Research is situated at

(a) Rajahmundry (b) Lucknow
(c) Hyderabad (d) Indore

128. Indian Institute of Pulses Research is situated at

(a) Karnal (b) Kanpur
(c) Junagadh (d) New Delhi

129. ICAR Research Complex for NEH Region is situated at

(a) Patna (b) Pune
(c) Meghalaya (d) New Delhi

130. Indian Agricultural Statistics Institute is situated at

(a) Bhopal (b) Kanpur
(c) Meghalaya (d) New Delhi

131. Indian Institute of Horticulture Research is situated at

(a) Bangalore (b) Patna
(c) Varanasi (d) Solan

132. Indian Lac Research Institute is situated at

(a) Bangalore (b) Ranchi
(c) Calcutta (d) Solan

133. Jute Technological Research Laboratory is situated at

(a) Bangalore (b) Ranchi
(c) Calcutta (d) Solan

134. National Centre for Integrated Pest Management is situated at

(a) Varanasi (b) Bhubaneshwar
(c) Faridabad (d) Cuttack

135. National Research Centre for Sorghum is situated at

(a) Solan (b) Hyderabad
(c) Karnal (d) None of these

136. National Centre for Weed Science is situated at

(a) Pune (b) Junagadh
(c) Lucknow (d) Jabalpur

137. National Bureau of Plant Genetic Resources is situated at

(a) Cuttack (b) Lucknow
(c) Nagpur (d) New Delhi

138. National Research Centre for Seed spices is situated at

(a) Calcutta (b) Ajmer
(c) New Delhi (d) None of these

139. Central Institute of Arid Horticulture is situated at

(a) Bikaner (b) Karnal
(c) Solan (d) Jhansi

140. Which is not a primary resource for agriculture

(a) Forest (b) Land
(c) Air (d) Fertilizer

141. Primary resource for agriculture is a

(a) Seed (b) Land
(c) Fertilizer (d) Air

142. Primary resource for agriculture is a

(a) Manures (b) Water
(c) Fertilizer (d) Pesticide

143. The environment of crop constitutes of

(a) Soil (b) Aerial
(c) Both (a) and (b) (d) None of these

144. The environment of crop amenable to modification

(a) Soil (b) Aerial
(c) Both (a) and (b) (d) None of these

145. Soil physical environment is

(a) Soil pH (b) Soil temperature
(c) Electrical conductivity (d) All of these

146. In the following soil chemical environment is

(a) Soil water (b) Soil temperature
(c) Electrical conductivity (d) All of these

147. Soil physical environment is

(a) Soil pH (b) Soil air
(c) Electrical conductivity (d) All of these

148. Which of the following is not a soil physical environment ?

(a) Soil nutrient (b) Soil air
(c) Soil temperature (d) Soil water

149. Which is a soil chemical environment ?

(a) Soil nutrient (b) Soil air
(c) Soil temperature (d) Soil water

150. In which BC cultivation of Potato begins ?

(a) 8800 BC (b) 3500 BC
(c) 7800 BC (d) 4700 BC

151. In which BC cultivation of Chickpea and Cotton begins ?

(a) 8000 BC (b) 2300 BC
(c) 7800 BC (d) 5200 BC

152. In which BC cultivation of Rice begins ?

(a) 7500 BC (b) 2200 BC
(c) 7800 BC (d) 6700 BC

153. In which BC cultivation of Ragi begins ?

(a) 8700 BC (b) 1800 BC
(c) 7700 BC (d) 5600 BC

154. In which BC cultivation of Sorghum begins ?

(a) 8800 BC (b) 1725 BC
(c) 7800 BC (d) 4600 BC

155. In which BC Wheel was invented ?

(a) 8700 BC (b) 3400 BC
(c) 7700 BC (d) 5700 BC

156. Who initiated experiments pertaining to plant nutrition in systematic way ?

(a) Jethro Tull (b) Gregor Johann Mendel
(c) Van Helmont (d) E.J. Butler

157. Department of Food was created in the year

(a) 1964 (b) 1942
(c) 1924 (d) 1956

158. AICRP on Maize was started in the year

(a) 1957 (b) 1940
(c) 1950 (d) 1972

159. AICRP on Dryland Agricultre was started in the year

(a) 1972 (b) 1950
(c) 1965 (d) 1974

160. NDP was started in the year

(a) 1950 (b) 1965
(c) 1988 (d) 1960

161. ________ Primitive form of agriculture is

(a) Shifting agriculture (b) Jhuming
(c) Hunting and Gathering (d) All of these

162. In India cereals occupy nearly ________ of cultivated lands

(a) 1/4 (b) 1/2
(c) 2/3 (d) 1/3

163. Who initiated the synthetic fertilizer industry ?

(a) Louis Pasteur (b) Gregor Johann Mendel
(c) Jethro Tull (d) Lawes

164. Who patented the process of treating rock phosphate to produce superphosphate

(a) Yuang Long Ping (b) M.S. Randhawa
(c) Van Helmont (d) Lawes

165. The laws of heredity were discovered by

(a) E.J. Butler (b) Gregor Johann Mendel
(c) Van Helmont (d) K.C. Mehta

166. Imperial Agricultural Research Institute was established in the year

(a) 1805 (b) 1915
(c) 1924 (d) 1905

167. "Horse Hoeing Husbandry" book was written by

(a) N.A. Cobb (b) Oscar Brefeld
(c) Jethro Tull (d) R.A. Fisher

168. Who concluded that the main 'principle of vegetation' is water

(a) E.J. Butler (b) Gregor Johann Mendel
(c) Van Helmont (d) K.C. Mehta

169. The "Annals of Agriculture" was written by

(a) N.A. Cobb
(b) Aurthur Young
(c) R.A. Fisher
(d) Jethro Tull

170. The book "Elements of Agricultural Chemistry" was written by

(a) Louis Pasteur
(b) Sir Humphry Davy
(c) Anton Debary
(d) Robert Haring

171. Experiments on effects of manures on crop were initiated by

(a) Sir John Bannet Lawes
(b) Justus von Liebig
(c) Oscar Brefeld
(d) R.A. Fisher

172. The rice crop tolerates submergence because __________.

(a) The plant has mechanism to transport O_2 through leaves
(b) The Plant does not need O_2
(c) O_2 penetrating to root zone through air movement is sufficient
(d) None of these

Fill in the Blanks

1. The term "Green Revolution" coined by __________.
2. The core metal in chlorophyll is __________.
3. In India, first separate Department of Agriculture was established in the year__________.
4. The word 'agros' means ________.
5. The first agricultural university in India is _________.
6. Central Rice Research Institute is located at _________.
7. National Bureau of Plant Genetics Resources is situated at ___________.
8. Dr.V. Kurian is associated with _____________ .
9. The term "Photoblasty" refers to the response of the _________ .
10. Agriculture Price Commission was set up in _____ .
11. Name the pathogen responsible for the famine in 1845 is _________.
12. The term "Operational holding" refers to the ________.
13. The sowing method that ensures uniform plant population is _________.
14. 'Krishi Pandit' is a title awarded every year by __________ .
15. Peter Dearesenzi is named as father of agronomy for his work on ________.
16. State with highest area under salt affected soil is _________ .
17. The Government of India decided to conduct Agricultural Census from the year ____________ .

18. Central Soil Salinity Research Institute is situated at _________ .
19. Hydroponic is cultivation in _________ .
20. Sunflower oil is rated as good quality oil because it contains ___________.

True/False

1. In Bt. Cotton the main purpose of breeding through genetic engineering is for pest resistance.**(T)**
2. Grow more food Campaign was started in 1953.**(F)**
3. Indian Council of Agricultural Research was set up in 1929.**(T)**
4. Truthful seed is used to produce foundation seed.**(F)**
5. The important feature of shifting cultivation is rotation of fields.**(T)**
6. Growing two or more crops simultaneously in the same field is inter-cropping.**(T)**
7. Growing of crops one after other to maintain soil fertility is known as crop rotation.**(T)**
8. The crops which are grown during summer are known as kharif.**(F)**
9. A farming system in which arable crops are grown with trees or shrubs established mainly to hasten soil fertility restoration and enhance soil productivity is called alley cropping.**(T)**
10. A short duration crop in between two main crops is termed as cash crop.**(F)**
11. Trap crops refer to crop grown around the main crop which are more susceptible to insect pest and diseases than the main crop.**(T)**
12. Growing four crops sequentially in a year in the same field is double cropping.**(F)**
13. In relay cropping second crop is sown before the harvest of previous crop.**(T)**
14. In mixed farming the main objective is higher profitability.**(F)**
15. In multistories cropping the principal of Annidation in space is used.**(T)**

Answers Fill in the Blanks

1. Dr. William Gaud
2. Mg
3. 1881
4. Field
5. G.B. Pant University of Agriculture and Technology, Pantnagar, Uttarakhand
6. Cuttack
7. New Delhi
8. White revolution
9. Seedling to light
10. 1965
11. *Phytophtora infestans*
12. Netland area cultivated by a farmer
13. Dibbling

14. ICAR
15. Literature collection of agronomy
16. Gujarat
17. 1970-71
18. Karnal
19. Water
20. High quantities of unsaturated fatty acids

ANSWERS

1.(d)	2.(d)	3.(a)	4.(c)	5.(a)	6.(c)	7.(c)
8.(a)	9.(d)	10.(b)	11.(d)	12.(c)	13.(c)	14.(b)
15.(c)	16.(c)	17.(d)	18.(b)	19.(a)	20.(c)	21.(a)
22.(c)	23.(a)	24.(a)	25.(d)	26.(b)	27.(a)	28.(c)
29.(a)	30.(b)	31.(b)	32.(b)	33.(b)	34.(c)	35.(c)
36.(d)	37.(b)	38.(b)	39.(b)	40.(d)	41.(b)	42.(b)
43.(a)	44.(a)	45.(b)	46.(c)	47.(d)	48.(b)	49.(a)
50.(d)	51.(b)	52.(d)	53.(d)	54.(b)	55.(b)	56.(d)
57.(a)	58.(a)	59.(d)	60.(d)	61.(c)	62.(b)	63.(b)
64.(a)	65.(a)	66.(d)	67.(a)	68.(d)	69.(d)	70.(b)
71.(d)	72.(d)	73.(b)	74.(a)	75.(a)	76.(a)	77.(b)
78.(c)	79.(d)	80.(b)	81.(b)	82.(c)	83.(d)	84.(a)
85.(d)	86.(d)	87.(c)	88.(b)	89.(a)	90.(a)	91.(b)
92.(c)	93.(d)	94.(a)	95.(b)	96.(a)	97.(b)	98.(b)
99.(c)	100.(b)	101.(b)	102.(b)	103.(a)	104.(c)	105.(a)
106.(d)	107.(a)	108.(b)	109.(d)	110.(a)	111.(a)	112.(c)
113.(d)	114.(a)	115.(b)	116.(a)	117.(a)	118.(a)	119.(a)
120.(a)	121.(d)	122.(a)	123.(b)	124.(c)	125.(d)	126.(c)
127.(c)	128.(b)	129.(c)	130.(c)	131.(a)	132.(b)	133.(c)
134.(c)	135.(b)	136.(d)	137.(d)	138.(b)	139.(a)	140.(a)
141.(b)	142.(b)	143.(c)	144.(c)	145.(b)	146.(c)	147.(b)
148.(a)	149.(a)	150.(b)	151.(b)	152.(b)	153.(b)	154.(b)
155.(b)	156.(c)	157.(b)	158.(a)	159.(a)	160.(c)	161.(d)
162.(c)	163.(d)	164.(d)	165.(b)	166.(d)	167.(c)	168.(c)
169.(b)	170.(b)	171.(a)	172.(a)			

2

Agricultural Meteorology

1. **The word "Meteorology" is derived from**

   (a) Greek language (b) Latin language
   (c) English language (d) French language

2. **International Meteorological Organization (IMO) was established in the year**

   (a) 1850 (b) 1932
   (c) 1873 (d) 1960

3. **Principles of mercury barometer was discovered by**

   (a) Paul Muller (b) Aristotle
   (c) Pliny (d) Torricelli

4. **Agricultural Meteorology Division at Pune was started in the year**

   (a) 1910 (b) 1930
   (c) 1932 (d) 1940

5. **High temperature in summer alternating with low temperature in winter is a specific condition of**

   (a) Tropical (b) Polar
   (c) Sub-tropical (d) All of these

6. **Which of the following colour absorbs more radiation ?**

   (a) Green (b) Black
   (c) Blue (d) White

7. **Blue colour of the sky is due to**

   (a) Reflection (b) Refraction
   (c) Dispersion (d) Radiation

8. **99 per cent of total mass of atmosphere is within _____ km.**
   (a) 30 (b) 10
   (c) 25 (d) 40

9. **Optimum temperature for wheat at vegetative, reproductive and grain development stages is**
   (a) 20, 18 and 10° C (b) 18, 20 and 10° C
   (c) 10, 20 and 18° C (d) 20, 30 and 31° C

10. **Optimum temperature for rice at tillering, flowering and ripening stages is**
   (a) 30, 20 and 31° C (b) 31, 30 and 20° C
   (c) 30, 20 and 20° C (d) 20, 30 and 31° C

11. **Most sensitive stage for low temperature in rice is**
   (a) Young microspore (b) Tillering
   (c) Seedling (d) Flowering

12. **For higher yields in crops, which among those must be maximum at flowering ?**
   (a) LAI and CGR (b) LAI and RGR
   (c) LAI and NAR (d) NAR and CGR

13. **Relationship between the amount of heat and plant development was quantified by**
   (a) Gregory (b) Mendel
   (c) Watson (d) Reaumur

14. **In which of the following condition, the sky radiation is more ?**
   (a) High latitudes (b) Winter season
   (c) Cloudy day (d) All of these

15. **Most of the tropical plants are**
   (a) C_3 Plants (b) CAM Plants
   (c) C_4 Plants (d) All of these

16. **Which among the plants have the highest water use efficiency ?**
   (a) CAM Plants (b) C_3 plants
   (c) C_4 Plants (d) None of these

17. **Photorespiration is higher in which of the following plant ?**
   (a) C_3 Plants (b) C_4 plants
   (c) CAM Plants (d) None of these

18. **"Climate" word derived from**
   (a) Latin language (b) Greek language
   (c) English language (d) French language

19. Statistical method of Forecasting is useful for preparation of

(a) Short range forecast (b) Long range forecast
(c) Both (a) and (b) (d) None of these

20. Criteria considered by planning commission to classify the agro climatic zones in India are

(a) Rainfall and temperature
(b) Soil, topography and water resources
(c) Farming and cropping systems
(d) All of these

21. In atmospheric layers, homosphere means

(a) Troposphere (b) Stratosphere
(c) Both (a) and (b) (d) Hydrosphere

22. Which of the following layer of atmosphere is called as hetrosphere

(a) Mesosphere (b) Stratosphere
(c) Thermosphere (d) Both (a) and (c)

23. Which is a short wave radiation in solar spectrum ?

(a) Visible light rays (b) UV rays
(c) Both (a) and (b) (d) None of these

24. Which of the rays constitutes major amount of solar radiation in solar spectrum ?

(a) Visible range (b) UV range
(c) Infrared range (d) None of these

25. Average pressure is equal to

(a) $10.34\ g/cm.s^2$ (b) $109.5\ g/cm.s^2$
(c) $103.4\ g/cm.s^2$ (d) $1040\ g/cm.s^2$

26. Normal sea surface temperature is

(a) 23° C (b) 28° C
(c) 18° C (d) 32° C

27. Geostationary satellites are left in earth's orbit at a height of ___ km.

(a) 860 (b) 8600
(c) 36000 (d) 360000

28. Heat transfer process by the molecular activity is known as

(a) Conduction (b) Convection
(c) Radiation (d) None of these

29. Abrupt rise of temperature instead of fall in temperature with increasing altitude is known as

(a) Lapse rate (b) Adiabatic lapse rate
(c) Inversion (d) None of these

30. Heat transfer in soil is mainly due to _____ process.

(a) Conduction (b) Convection
(c) Radiation (d) Global warming

31. In soil heat flow is mainly depends on _______

(a) Temperature gradient (b) Thermal conductivity
(c) Both (a) and (b) (d) None of these

32. In soil thermal conductivity is depends on

(a) Texture (b) Moisture
(c) Organic matter (d) All of these

33. Maximum diurnal variation of atmosphere is ?

(a) 16 (b) 25
(c) 24 (d) 36

34. Approximate surface temperature of sun is

(a) 300° C (b) 8000° C
(c) 6000° C (d) 7000° C

35. Region between tropic of cancer (23 1/2° N) and the tropic of capricon (23 1/2° S) is

(a) Polar zone (b) Temperature zone
(c) Subtropical zone (d) Tropical zone

36. Region between tropic of cancer and the Artic circle (23 1/2 N to 66 1/2 N) and the tropic of capricon and Antarctic circle(23 1/2 S to 66 1/2° S) is

(a) Polar zone (b) Temperature zone
(c) Subtropical zone (d) Tropical zone

37. Force which tends to turn the flow of air due to earth's rotation is

(a) Land breeze (b) Sea breeze
(c) Coriolis force (d) None of these

38. Which stage is highly sensitive to high temperature ?

(a) Flowering stage (b) Vegetative stage
(c) Seedling stage (d) Maturity stage

39. Optimum photo-period requirement of rice at flowering stage is ?

(a) 7.0 to 8.0 hrs (b) 9.0 to 10 hrs
(c) 10.0 to 10.5 hrs (d) 11.0 to 13.0 hrs

40. RH is measured continuously by

(a) Hygrograph (b) Hair hygrometer
(c) Assmanns psychrometer (d) None of these

41. RH increases pest and disease incidence

(a) Decreases (b) Increases
(c) Increases and decreases (d) None of these

42. Range of RH conducive for crop production is

(a) 50-60% (b) 60-80%
(c) 80-100% (d) 30-40%

43. At higher RH, temperature will be

(a) Low (b) High
(c) Initially low and increases (d) Initially high and decreases

44. RH is the highest at

(a) Poles (b) Equator
(c) Horse latitude (d) All of these

45. It is a process by which water in a liquid state is converted to vapour is called

(a) Transpiration (b) Evaporation
(c) Evapotranspiration (d) Sublimation

46. Conversion efficiency (calories required for production of one gram dry matter) in crop plants ranges between ______ cal/g.

(a) 1000 and 1800 (b) 1500 and 3000
(c) 2700 and 4100 (d) 4200 and 5500

47. Which among the operation increase the utilization of solar energy

(a) Irrigation (b) Nitrogen application
(c) High plant population (d) All the above

48. Which among the following is a condensation form of precipitation ?

(a) Fog (b) Dew
(c) Fost (d) All of these

49. Due to the mountain barrier, rain occurs, is a process of

(a) Convection process
(b) Orographic process
(c) Cyclone
(d) None of these

50. Snow cover keeps the soil

(a) Hot (b) Cool
(c) Both (a) and (b) (d) None of these

51. Which of the following is used for knowing the increasing or decreasing trend of rainfall ?

(a) Periodicity of rainfall (b) Weekly rainfall
(c) Monthly rainfall (d) Decennial rainfall

52. Doldrums (Low pressure belt at equator) are confined to

(a) 5° N and S for equator (b) 10° N and S of equator
(c) 15° N and S of equator (d) None of these

53. For every 1000 feet height, pressure decreases by _______ inch in the air.

(a) 1 (b) 4
(c) 2 (d) 5

54. During summer which among the area has higher pressure ?

(a) Land mass (b) Sea
(c) Forest (d) None of these

55. Greater role in heat balance of earth is________ .

(a) Temperature (b) Soil moisture
(c) Wind (d) None of these

56. During night hours wind moves from mountain to valley is known as

(a) Land breeze (b) Sea breeze
(c) Mountain breeze (d) Valley breeze

57. For winnowing operation required wind speed is

(a) 20 km/hr (b) 25 km/hr
(c) 15 km/hr (d) 10 km/hr

58. Plant growth generally inhibited if the wind velocity is more than

(a) 5 km/hr (b) 10 km/hr
(c) 15 km/hr (d) 20 km/hr

59. Which among the following is having the least albedo ?

(a) Dry black soil (b) Moist black soil
(c) Grey soil (d) Moist grey soil

60. Photosynthetically active part of region is

(a) 0.3- 0.4 and 0.5-0.6 micron (b) 0.4-0.5 and 0.7-0.8 micron
(c) 0.4- 0.5 and 0.6-0.7 micron (d) None of these

61. Light intensity at which maximum rate of photosynthesis is obtained is called

(a) Light compensation point (b) Light saturation point
(c) Both (a) and (b) (d) None of these

62. Cloud which is commonly forming hallow around sun is known as

(a) Cumulonimbus (b) Cirrostratus
(c) Cumulus (d) None of these

63. Which cloud responsible for continuous precipitation ?

(a) Nimbostratus (b) Cirrus
(c) Cumulus (d) Cumulonimbus

64. Generally higher rainfall is received in

(a) 10° N and S of equator (b) 10-20° N and S of equator
(c) 20-30° N and S of equator (d) All of these

65. Cyclonic effect on agriculture is more common in India during the month of ______

(a) March-May (b) October-December
(c) January-February (d) June-September

66. According to India Meteorological Department, range of rainfall variation is considered as normal is

(a) -39 to +39% (b) -19 to +19%
(c) -29 to +29% (d) None of these

67. For calculating initial probability, ______ is taken as optimum

(a) 40% (b) 50%
(c) 80% (d) 28%

68. Which among the following light is most favourable for growth of plants ?

(a) Green (b) White
(c) Red (d) Yellow

69. Solar radiation for rice is crucial at

(a) Panicle initiation stage (b) Tillering stage
(c) Flowering stage (d) Panicle initiation to flowering

70. Rice yields are higher during dry season is mainly due to higher

(a) Solar radiation (b) Temperature
(c) Wind (d) None of these

71. In cereals, low light intensity is essential during which one of the following stage

(a) Germination (b) Tillering
(c) Grain formation (d) Flowering to grain formation

72. **Which among the following crop is having constant harvest index ?**
(a) Rice (b) Groundnut
(c) Maize (d) Pearl millet

73. **Specific heat of water, soil and organic matter is**
(a) 0.2, 1.0 and 0.4 (b) 1.0, 0.2 and 0.5
(c) 1.0, 0.5 and 0.3 (d) None of these

74. **Heat (calories) required to raise a unit mass (g) of substance by one degree Celsius is known as**
(a) Heat capacity (b) Specific heat
(c) Thermal diffusivity (d) None of these

75. **Optimum temperature for nitrification process is**
(a) 10-15° C (b) 35-40° C
(c) 25-30° C (d) None of these

76. **Optimum temperature for nodule formation is**
(a) 24° C (b) 14° C
(c) 18° C (d) 34° C

77. **Which among the following parameter is more important in weather forecasting ?**
(a) Sunlight (b) Wind
(c) Temperature (d) Atmospheric pressure

78. **Atmospheric pressure is measured by which one of the following ?**
(a) Psycrometer (b) Thermograph
(c) Barograph (d) Cismograph

79. **Absolute zero is equal to**
(a) 273° C (b) 273° F
(c) -273° C (d) 273° K

80. **Centigrade can be converted to Kelvin by**
(a) Deducting 100 (b) Adding 273
(c) Adding -273 (d) None of these

81. **Liquid used in maximum and minimum thermometer, respectively are**
(a) Alcohol and Mercury (b) Mercury and Alcohol
(c) Mercury and Lead (d) Alcohol and Lead

82. **Air temperature is continuously measured by which of the following**
(a) Maximum thermometer (b) Minimum thermometer
(c) Thermograph (d) Mercury thermometer

83. Optimum temperature range in higher plants for its maximum dry matter production is

(a) 0-25° C (b) 25-50° C
(c) 10-40° C (d) 30-50° C

84. The average value of solar constant is

(a) 1.94 J/cm²/min (b) 1.94 cal/cm²/min
(c) 2.94 J/cm²/min (d) 2.94 cal/cm²/min

85. Which type of radiation is useful in remote sensing is

(a) Diffused radiation (b) Direct radiation
(c) Reflected radiation (d) Net radiation

86. Radiation balance between global and reflected solar radiation is known as

(a) Diffused radiation (b) Albedo
(c) Net radiation (d) None of these

87. Red colour of sunset is due to

(a) Dispersion (b) Reflection
(c) Refraction (d) None of these

88. Average absorption of solar radiation by the earth surface is

(a) 34% (b) 10%
(c) 40% (d) 15%

89. Terrestrial radiation is transferred upward as

(a) Long wave radiation (b) Short wave radiation
(c) Both (a) and (b) (d) None of these

90. Earth rotates in its axis at which angle ?

(a) 30 ½ (b) 42 ½
(c) 18 ½ (d) 23 ½

91. In northern hemisphere shortest day is on

(a) March 31 (b) May 31
(c) September 23 (d) December 21

92. Both day and night are equal on

(a) March 21 (b) September 23
(c) Both (a) and (b) (d) None of these

93. Visible radiation is the range of

(a) 200 to 600 micron (b) 490 to 700 micron
(c) 390 to 760 micron (d) 790 to 870 micron

94. SI unit of solar radiation is

(a) Watt/m^2 (b) Watt/m
(c) Joule/m (d) None of these

95. Photosynthetically active radiation can be measured in ______ units

(a) Photon (b) Watt
(c) Einstein (d) ppm

96. Solid portion of the earth is

(a) Lithosphere (b) Hydrosphere
(c) Atmosphere (d) Ionosphere

97. Hydrosphere is earth's

(a) Solid portion (b) Water portion
(c) Gaseous portion (d) All of these

98. Micro-climate is synonym of

(a) Plant climate (b) Eco-climate
(c) Habitat climate (d) All of these

99. Which among the following will decide the time of sowing in dry-land condition?

(a) Wind velocity (b) Rainfall
(c) Low temperature (d) High temperature

100. Water vapour content at the lower part of atmosphere is (by volume)

(a) 0-2% (b) 0-3%
(c) 0-6% (d) 0-4%

101. Transition zone between troposphere and stratosphere is known as

(a) Mesopause (b) Ionosphere
(c) Tropopause (d) None of these

102. Ozonosphere is a synonym of

(a) Hydrosphere (b) Mesosphere
(c) Thermosphere (d) Stratosphere

103. Atmospheric temperature variation in four layers is

(a) Go on decreases from troposphere to thermosphere
(b) Go on increases from troposphere to thermosphere
(c) Increases in troposphere and changes alternatively
(d) Decreases in troposphere and changes alternatively

104. Communication of Radio is made possible through

(a) Ionosphere (b) Hydrosphere
(c) Stratosphere (d) None of these

105. Distance between earth and sun is _____ km

(a) 1.5×10^8 (b) 1.8×10^7
(c) 1.7×10^9 (d) 1.5×10^6

106. Solar radiation is measured by which instrument ?

(a) Glucometer (b) Barograph
(c) Thermometer (d) Pyranometer

107. In most of the observatories, solar radiation is measured in India by

(a) Photochemical sunshine recorder (b) Cambell stokes sunshine recorder
(c) Bellany pyrenometer (d) None of these

108. The process responsible for heat flow is _____

(a) Conduction (b) Convection
(c) Radiation (d) All of these

109. Which among the following does not require medium for its heat flow ?

(a) Conduction (b) Convection
(c) Radiation (d) None of these

110. Maximum temperature in a day is observed at

(a) 11.00 am (b) 2.00 pm
(c) 5.00 pm (d) 1.00 noon

111. Day and night range at equator is

(a) More day than night (b) More night than day
(c) Both are equal (d) Less night than day

112. Photosensitive varieties are

(a) Period bound (b) Season bound
(c) Both (a) and (b) (d) None of these

113. Photoperiodism is used in crop production for

(a) Identify the suitable variety (b) Optimum date of sowing
(c) Both (a) and (b) (d) None of these

114. The share of Photosynthetically Active Radiation (PAR) in crop production is

(a) 4 -7% (b) 1-2%
(c) 10 -15% (d) 20 -25%

115. Carbon dioxide fixation in crop plants is about

(a) 15% (b) 40%
(c) 20% (d) 60%

116. For source limited crops, solar radiation is important during

(a) Panicle initiation stage (b) Flowering stage
(c) Tillering stage (d) Grain filling stage

117. Solar energy utilization is high during ____ of crop.

(a) Germination stage (b) Vegetative stage
(c) Flowering stage (d) Maturity stage

118. Hail and Hailstorms occur due to presence of

(a) Cumulus cloud (b) Nimbus cloud
(c) Cirrus cloud (d) Cumulonimbus cloud

119. When due point temperature is below 0°C, moisture turns directly from gaseous to solid state called

(a) Frost (b) Glane
(c) Sleet (d) None of these

120. Which among the following is Cloud Condensation Nucleus ?

(a) Dust (b) Smoke
(c) Salt particles (d) All of these

121. Warm rain (Rainfall in the tropics) is due to the process of

(a) Coalescence (b) Melting
(c) Sublimation (d) None of these

122. Which material is used in artificial rain making ?

(a) Dry ice (b) Silver iodide
(c) Common salt (d) All of these

123. In warm rain process, cloud seeding is made by

(a) Common salt (b) Dry ice
(c) Silver iodide (d) All of these

124. Which state having largest area under arid climate in India ?

(a) Bihar (b) Rajasthan
(c) Madhya Pradesh (d) Andhra Pradesh

125. The state which is having more area under semi-arid region is

(a) Maharashtra (b) Madhya Pradesh
(c) Karnataka (d) Andhra Pradesh

126. Planning commission divided India into different Agro-climatic zones during

(a) 1989 (b) 1985
(c) 1990 (d) 1970

127. Agro-climatic zones are formed in India during

(a) IV Five year plan (b) IX Five year plan
(c) VII Five year plan (d) X Five year plan

128. Shifting cultivation is more common in India is ____ zone.

(a) Western Himalayan (b) Eastern Himalayan
(c) Upper Gangetic (d) Both (a) and (c)

129. Best quality oranges are grown in

(a) Eastern Plateau and hills (b) Northern plateau and hills
(c) Western plateau and hills (d) Southern plateau and hills

130. Oilseeds zone in India is

(a) Gujarat plains and hills (b) Western plateau and hills
(c) Island zone (d) None of these

131. At what depth below the soil, temperature variation is nil ?

(a) 30 cm (b) 60 cm
(c) 80 cm (d) 40 cm

132. Absolute temperature is measured in

(a) Celsius (b) Fahrenheit
(c) Kelvin (d) All of these

133. Lines connecting the points of equal water table are known as

(a) Isotherm (b) Isobar
(c) Isobath (d) None of these

134. Commonly used instrument for RH measurement is

(a) Wet and dry bulb thermometers (b) pH meter
(c) Hygrograph (d) Seismograph

135. Millibar of pressure is equal to

(a) 100 N/m^2 (b) 200 N/m^2
(c) 500 N/m^2 (d) 600 N/m^2

136. One cubic meter air can hold _____ of water vapour at 10° C

(a) 8.0 g (b) 4.0 g
(c) 10.0 g (d) 11.4 g

137. Condensation takes place when the RH reaches ____

(a) 60% (b) 70%
(c) 80% (d) 100%

138. What character is considered for classifying cloud species?

(a) Shape (b) Internal structure
(c) Both (a) and (b) (d) None of these

139. Clouds which are classified under high level clouds ?

(a) Cirrus
(b) Cirrostratus
(c) Cirrocumuls
(d) All of these

140. Increase in altitude of a place makes

(a) Decrease in temperature
(b) Increase in precipitation
(c) Increase in wind velocity
(d) All of these

141. Which one of the following is the most important factor affects climate ?

(a) Latitude
(b) Distribution of land
(c) Ocean currents
(d) All of these

142. Which cloud responsible for rainfall during summer season ?

(a) Cumulonimbus
(b) Stratus
(c) Cumuls
(d) Stratocumulns

143. Which among the following is a rain bearing clouds ?

(a) Cirrus
(b) Stratus
(c) Nimbostratus
(d) Cumulus

144. Evaporimeter developed in Israel and used to measure evaporation for short period is

(a) USWB pan evaporimeter
(b) Colorado Sunken screen evaporimeter
(c) Can evaporimeter
(d) Portable evaporimeter

145. High and low pressures are developed mainly due to

(a) Amount of solar radiation received
(b) Absorption of heat by water and earth
(c) Rotation of earth
(d) All of these

146. During hot weather period (March-May) which of the conditions will prevail in India?

(a) Low pressure through around Thar desert
(b) Pre-monsoon showers over Kerala
(c) Dry hot winds in North West India
(d) All of these

147. In Eastern Asia and the Philippines, cyclones are known as

(a) Hurricane
(b) Typhoon
(c) Baquio
(d) T-sunami

148. Fall of pressure from the periphery to the centre of the storm ranged between

(a) 20 and 30 millibar
(b) 20 and 50 millibar
(c) 20 and 70 millibar
(d) 10 and 20 millibar

149. Torrential down pour occur at ____ from the centre of the cyclone

(a) 10-20 km
(b) 20-30 km
(c) 30-50 km
(d) 50-100 km

150. Short range forecasting is useful for

(a) Irrigation engineer
(b) Farmers/ Gardners
(c) Mariners
(d) All of these

151. In forecasting of rainfall, the word 'Scattered rains' denotes ____ area receives rain

(a) 1/4
(b) 1/3
(c) 2/3
(d) None of these

152. Satellite takes the pictures in

(a) Red region
(b) Infra red region
(c) Blue region
(d) None of these

153. It earth has no atmosphere, earth's temperature will be

(a) +18° C
(b) -18° C
(c) +28° C
(d) -24° C

154. Which among the following is green house gas ?

(a) CO_2
(b) Methane
(c) CFC s
(d) All of these

155. Carbon dioxide (CO_2) concentration is increasing at the rate of

(a) 0.04%/ day
(b) 0.04%/ week
(c) 0.04%/month
(d) 0.04%/ year

156. CO_2 concentration increases due to

(a) Deforestation
(b) Industrialization
(c) Human respiration
(d) None of these

157. Impact of green house is

(a) Global warming
(b) Sea level rise
(c) Ice formation
(d) Both (a) and (b)

158. Gas responsible for ozone layer depletion is

(a) CF_2Cl_2
(b) CO_2
(c) SO_2
(d) C_2H_2

159. Which among the following gas contributes about 50% of global warming?

(a) CO_2 (b) CFC
(c) CH_4 (d) N_2O

160. In Hargreaves "Moisture Availability Index" the percentage of dependable rainfall is

(a) 30% (b) 60%
(c) 75% (d) 90%

161. According to National Bureau of Soil Survey and Land Use Planning, the number of Agro-eco regions are

(a) 9 (b) 18
(c) 20 (d) 28

162. Long range weather forecasting is useful for

(a) Postponing irrigation (b) Choosing cropping pattern
(c) Pesticide application (d) Fertilizer application

163. Method of Synoptic forecasting is useful for

(a) Short range forecast (b) Medium range forecast
(c) Long range forecast (d) None of these

164. A line of equal wind speed is known as

(a) Isobar (b) Isotach
(c) Isobath (d) None of these

165. Solar constant is a measure of

(a) Temperature (b) Flux density
(c) Humidity (d) None of these

166. Normal lapse rate is

(a) 5.5° C/km (b) 8.5° C/km
(c) 6.5° C/km (d) 9.5° C/km

167. With increasing porosity and higher organic matter, thermal conductivity will

(a) Decrease and then increase (b) Decrease
(c) Increase and then decrease (d) None of these

168. Lowest temperature in a day is observed during

(a) 6.00 am (b) Just before sunrise
(c) Sunrise (d) None of these

169. Melting and Boiling points of water in F unit are

(a) 32° F and 210° F (b) 23° F and 98.4° F
(c) 38.9° F and 200° F (d) 24° F and 210° F

170. Melting point of water in Kelvin unit is

(a) -273 K (b) 273K
(c) 108 K (d) 38 K

171. Lifting up of plants along with soil due to very low temperature is

(a) Suffocation (b) Heaving
(c) Freezing injury (d) None of these

172. Thermal death point of crop plants is at which temperature ?

(a) 50-60° C (b) 20-30° C
(c) 30-40° C (d) 40-50° C

173. An injury known as stem griddle is caused by

(a) Low soil moisture (b) High soil temperature
(c) Low soil temperature (d) None of these

174. Daily variation of soil temperature is negligible at _____ depth of soil

(a) 12 cm (b) 15 cm
(c) 30 cm (d) 6 cm

175. Cardinal temperature is

(a) Maximum temperature (b) Minimum temperature
(c) Optimum temperature (d) All of these

176. Optimum temperature for cool season crops range between

(a) 15 and 20° C (b) 10 and 15° C
(c) 25 and 30° C (d) 35 and 40° C

177. Optimum temperature for warm season crops ranges between

(a) 30 and 38° C (b) 35 and 45° C
(c) 15 and 25° C (d) 20 and 30° C

178. The warm season crop is/are

(a) Rice (b) Sorghum
(c) Groundnut (d) All of these

179. Which among the stage requires lower temperature for higher productivity ?

(a) Flowering to maturity
(b) Seedling to panicle initiaion
(c) Panicle initiation to flowering
(d) None of these

180. Photothermal unit is the combination of the following

(a) Temperature and light (b) Temperature and RH
(c) Temperature and photo period (d) None of these

181. Prolonged low pressure centre is called as

(a) Trough (b) Bows
(c) Cyclones (d) None of these

182. Amount of water vapour in the atmosphere depends on

(a) Wind (b) Temperature
(c) Both (a) and (b) (d) None of these

183. Ratio between weight of water vapour and volume of air is

(a) Absolute humidity (b) Vapour Pressure Deficit (VPD)
(c) Specific humidity (d) None of these

184. Water vapour present in the air to water vapour required for saturation is

(a) Specific humidity (b) Relative humidity
(c) Vapour Pressure Deficit (VPD) (d) None of these

185. Low RH includes

(a) More evaporation (b) More transpiration
(c) Both (a) and (b) (d) None of these

186. RH is expressed as

(a) Percentage (b) Ratio
(c) km/hr (d) All of these

187. Stevenson's screen placed at a height of

(a) 4 feet (b) 8 feet
(c) 5 feet (d) 9 feet

188. Evaporation is given as

(a) mm/min (b) cm/day
(c) mm/day (d) None of these

189. Diameter and height of the USWB open pan evaporimeter is

(a) 110, 15 cm (b) 122, 25 cm
(c) 160, 20 cm (d) None of these

190. Evaporimeter used for estimating ET under field condition and scheduling of irrigation is

(a) Portable evaporimeter (b) Sunken screen evaporimeter
(c) Can evaporimeter (d) None of these

191. Wind effects on plant growth by

(a) Transpiration increase (b) Increase intake of CO_2
(c) Mechanical damage (d) All of these

192. Basic objective of shelter belt is

(a) Reduce wind velocity (b) Shade the crop
(c) Reduce the temperature (d) Increase the water table

193. Horse latitudes (High Pressure Belts) are confined to

(a) 0-10 N and S (b) 10-15 N and S
(c) 15-20 N and S (d) 30-35 N and S

194. Air moves from high pressure belt to equator is

(a) Trade wind (b) Anti-trade wind
(c) Both (a) and (b) (d) None of these

195. 'Monsoon' word derived from ________ language.

(a) Greek (b) Latin
(c) French (d) Arab

196. Normally, south west monsoon starts in Kerela at

(a) May 1st week (b) May 3rd week
(c) June 1st week (d) June 2nd week

197. Which among the state is benefited by North East monsoon ?

(a) Tamil Nadu (b) Andhra Pradesh
(c) Both (a) and (b) (d) None of these

198. Pre-monsoon dry sowing is done with predicting the rainfall by

(a) Conditional probability (b) Initial probability
(c) Both (a) and (b) (d) None of these

199. Which of the following is the example of weather ?

(a) Hot day (b) Rainy day
(c) Cloudy day (d) All of these

200. Which of the following is used to locate the thunderstorms ?

(a) Sonar (b) Radar
(c) Both (a) and (b) (d) None of these

201. Which among the following will resist the wind movement ?

(a) Wind breaks (b) Shelter belts
(c) Both (a) and (b) (d) None of these

202. By covering the soil with the mulches, soil temperature gets

(a) No change (b) Increase
(c) Decrease and then increase (d) Increase and then decrease

203. Cold waves and frost are more problematic in India is

(a) Northern India (b) Southern India
(c) Eastern India (d) Uncertain

204. Among the weather abnormalities, which is most destructive ?

(a) Frost (b) Drought
(c) Tornado (d) Cold waves

205. Generally cyclone moves

(a) Westward (b) North-westward
(c) South-westward (d) Both (a) and (b)

206. Average speed of cyclone is _____ km/h.

(a) 25 (b) 8
(c) 10 (d) 18

207. In night hours, wind moves from land to sea is known as

(a) Land breeze (b) Sea breeze
(c) Mountain breeze (d) Valley breeze

208. Highest per cent of cover is in

(a) Central plateau and hills (b) Western plateau and hills
(c) Island zone (d) Gujarat plains and hills

209. Number of Agro Ecological zones in India is

(a) 8 (b) 16
(c) 14 (d) 11

210. In forecasting of rainfall, the word "isolated rain" denotes ____ area receives rain

(a) 1/5 (b) 1/2
(c) 1/3 (d) None of these

211. Which of the following is the most important weather elements ?

(a) Temperature and Precipitation (b) Temperature and RH
(c) RH and Wind (d) Precipitation and wind

212. AICRPDA classified the climate based on

(a) Moisture Deficit index (b) Moisture Availability Index
(c) Both (a) and (b) (d) None of these

213. Thornthwaite classified the climate based on

(a) PE (Precipitation Effectiveness) index
(b) TE (Temperature effectiveness) index
(c) Seasonal distribution of rainfall
(d) All of the above

214. The climate classified by Koppen is based on

(a) Temperature (b) Precipitation
(c) Vegetation (d) Both (a) and (b)

215. Straight card is used in sunshine recorder during

(a) 1st March to 11th April
(b) 1st September to 12th October
(c) 12th April to 31st August
(d) Both (a) and (b)

216. Long curved cards used in sunshine recorder during

(a) 1st March to 11th April
(b) 1st September to 12th October
(c) 13th October to end of February
(d) 12th April to 31st August

217. Which among the following is short day plant ?

(a) Soybean
(b) Mustard
(c) Sorghum
(d) All of these

218. Rice perform well at flowering with RH of

(a) 60-70%
(b) 70-80%
(c) 80-90%
(d) 50-60%

219. 1° longitude equals the time span of

(a) 2 minutes
(b) 3 minutes
(c) 5 minutes
(d) 4 minutes

220. In India rainfall is generally measured at _____ .

(a) 7.00 am
(b) 8.00 am
(c) 8.30 am
(d) 3.00 pm

221. Knots (Nautical miles per hour) is a unit of _____ measurement.

(a) pH
(b) Wind velocity
(c) Temperature
(d) Sound

222. Wet and dry bulb thermometers are used to measure

(a) Temperature
(b) pH
(c) RH
(d) Solar radiation

Fill in the blanks

1. The instrument used for recording the atmospheric humidity continuously is _____ .
2. Ozone layer absorbs ________ radiation.
3. Dry atmospheric air contains ________ % of carbon dioxide on volume basis.
4. Rainfall associated with the hailstones is called ________ .
5. Line of map which joins places of equal temperature is called ________ .
6. ________ clouds produce continuous precipitation.
7. The planet near earth is ________ .
8. ________ is used for measurement of evaporation rates in the field crop.

9. Saturation of atmospheric vapour forms __________ .
10. Acid rain water contains _______ and ______ .
11. _________ instrument is used for measuring all weather parameters.
12. Rainy day is considered when rainfall is above ______ mm.
13. World Meteorological observatory is located at _______ .
14. For installation of agro meteorology ________ size (L x B) is recommended.
15. Atmospheric pressure is measured with the help of ______ instrument.
16. _________ instrument used for measuring effective rainfall.
17. Thornthwaite climatic classification (1948) is revised by _______ .
18. Part of total rainfall used for crop production is termed as ________ rainfall.
19. Meteorological drought occurs when actual rainfalls below _______ level of normal rainfall.
20. Agro meteorology is the branch of meteorology which investigates the relationship of _________ and __________ to physical environment.

True/False

1. The first meteorological department of India was started at Shimla (1864). **(T)**
2. World Meteorological Organization (WMO) was established at U.S.A .**(F)**
3. Climate word originated from Latin.**(F)**
4. The study of physical status of climate is known as statistical climatology. **(F)**
5. The weather information collected from observatory are depicted in tables is called weather chart. **(T)**
6. To study the different factors of climate such as temperature and rainfall, are divided into state level is called Descriptive climatology.**(T)**
7. The science of climatological condition which have direct relation or relevance to agriculture is called Agro-meteorology.**(T)**
8. The composite or generalization of the variety of day to day weather condition is called season.**(T)**
9. When weather data collected from sloppy land is called topo climate.**(T)**
10. When weather data collected from standing crop is known as micro-climate.**(T)**
11. Maximum temperature in a day is generally observed at 11 am. **(F)**
12. Solar constant is a measure of flux density.**(F)**
13. Snow cover keeps the soil hot.**(T)**
14. The word 'Nimbus' is related with rain.**(T)**
15. Cyclones will form when the sea surface temperature exceeds 22.2° C. **(T)**

16. Satellite takes the pictures in yellow region.**(F)**
17. Knots (Nautical miles per hour) is a unit of wind velocity measurements.**(T)**
18. Radar used to locate thunderstorms.**(T)**
19. By covering the soil with the mulches, soil temperature gets increase. **(T)**
20. Albedo also called as diffused radiation. **(F)**

Answers Fill in the Blanks

1. Hygrograph
2. Ultraviolet
3. 0.03%
4. Hailstorm
5. Isotherm
6. Primary/secondary
7. Venus
8. Atmometer
9. Dew
10. Nitric acid and sulphuric acid
11. Auto Weather station
12. 2.5
13. Genewa Switzerland
14. 55 x 36m
15. Barometer
16. Ramdas
17. T. & Mather, 1955
18. Effective
19. 75 mm
20. Plant & animal

ANSWERS

1.(a)	2.(c)	3.(d)	4.(c)	5.(a)	6.(b)	7.(c)
8.(d)	9.(d)	10.(b)	11.(a)	12.(c)	13.(d)	14.(d)
15.(c)	16.(a)	17.(a)	18.(b)	19.(c)	20.(d)	21.(c)
22.(d)	23.(c)	24.(c)	25.(c)	26.(a)	27.(c)	28.(a)
29.(c)	30.(a)	31.(c)	32.(d)	33.(a)	34.(c)	35.(d)
36.(b)	37.(c)	38.(c)	39.(c)	40.(a)	41.(b)	42.(b)
43.(a)	44.(b)	45.(b)	46.(c)	47.(d)	48.(d)	49.(b)
50.(a)	51.(d)	52.(a)	53.(a)	54.(b)	55.(c)	56.(c)
57.(c)	58.(b)	59.(b)	60.(c)	61.(b)	62.(b)	63.(a)
64.(a)	65.(b)	66.(b)	67.(b)	68.(c)	69.(d)	70.(a)
71.(d)	72.(d)	73.(b)	74.(b)	75.(c)	76.(a)	77.(d)
78.(c)	79.(c)	80.(b)	81.(b)	82.(c)	83.(c)	84.(b)
85.(c)	86.(c)	87.(a)	88.(a)	89.(a)	90.(d)	91.(d)
92.(c)	93.(c)	94.(a)	95.(c)	96.(a)	97.(b)	98.(d)
99.(b)	100.(d)	101.(c)	102.(d)	103.(d)	104.(a)	105.(a)
106.(d)	107.(b)	108.(d)	109.(c)	110.(b)	111.(c)	112.(b)
113.(a)	114.(b)	115.(c)	116.(d)	117.(c)	118.(d)	119.(a)
120.(d)	121.(a)	122.(d)	123.(a)	124.(b)	125.(a)	126.(a)
127.(c)	128.(b)	129.(c)	130.(a)	131.(c)	132.(a)	133.(c)
134.(a)	135.(a)	136.(d)	137.(d)	138.(c)	139.(d)	140.(d)
141.(a)	142.(a)	143.(c)	144.(d)	145.(d)	146.(d)	147.(b)
148.(c)	149.(c)	150.(d)	151.(c)	152.(b)	153.(b)	154.(d)
155.(d)	156.(c)	157.(b)	158.(a)	159.(a)	160.(c)	161.(c)
162.(b)	163.(a)	164.(b)	165.(b)	166.(c)	167.(b)	168.(b)
169.(a)	170.(b)	171.(b)	172.(a)	173.(b)	174.(c)	175.(d)
176.(c)	177.(a)	178.(d)	179.(c)	180.(c)	181.(a)	182.(c)
183.(a)	184.(b)	185.(c)	186.(a)	187.(c)	188.(c)	189.(b)
190.(c)	191.(d)	192.(a)	193.(d)	194.(a)	195.(d)	196.(c)
197.(c)	198.(a)	199.(d)	200.(b)	201.(c)	202.(b)	203.(a)
204.(b)	205.(d)	206.(d)	207.(a)	208.(c)	209.(a)	210.(c)
211.(a)	212.(a)	213.(d)	214.(d)	215.(d)	216.(d)	217.(d)
218.(b)	219.(d)	220.(c)	221.(b)	222.(c)		

3

Soil Properties and Tillage

1. **It has been observed that soil pH increases in**
   (a) Pre-monsoon (b) Zaid
   (c) Winter (d) Post monsoon
2. **Which of the following soils have poor physical condition ?**
   (a) Saline (b) Sodic
   (c) Saline-alkali (d) Alkaline
3. **CEC values for organic soils lies in range of**
   (a) 50-100 meq/100g (b) 10-20 meq/100g
   (c) 20-50 meq/100g (d) 150-200 meq/100g
4. **Which extract is used for determination of CEC ?**
   (a) 1 N NH_4 OAC (b) 1 N $NaOHCO_3$
   (c) 1 N KOAC (d) Olsen reagent
5. **Which of the following formula is used for calculation of base saturation**
   (a) Total bases/ CEC × 100
   (b) Exchangeable bases / CEC × 100
   (c) CEC/ total bases × 100
   (d) CEC / exchange able bases ×100
6. **Red colour of the soil is due to presence of**
   (a) Limonite mineral (b) Hematite mineral
   (c) Turgite mineral (d) Geothite mineral
7. **Relative proportion of soil particles is known as**
   (a) Soil texture (b) Soil structure
   (c) Porosity (d) None of these

8. **Which of the following character doesn't exist in sand particles ?**
   (a) Plasticity (b) Cohesion
   (c) Stickiness (d) All of the these

9. **Clay particles exhibit which of the following character ?**
   (a) Flocculation (b) De-flocculation
   (c) Plasticity (d) All of the these

10. **Sandy soil has approx._____ percentage of sand**
   (a) 60 (b) 50
   (c) 80 (d) 99

11. **Humus in soil sample is removed by ______ chemical in mechanical analysis**
   (a) $KMnO_4$ (b) Hydrogen peroxide (H_2O_2)
   (c) Perchloric acid (d) HCl

12. **When sand particles are not coated with clay or silt, they do not exhibit**
   (a) Stickiness (b) Plasticity
   (c) Colloidal property (d) All of these

13. **Which of the following minerals (other than quartz), exist in sand and silt particles ?**
   (a) Feldspar (b) Mica
   (c) Hematite/ Limonite (d) All of these

14. **Which soil structure hinders free drainage ?**
   (a) Platy (b) Granular
   (c) Crumby (d) Columnar

15. **Soil condition exhibited at higher bulk density soil is**
   (a) More compaction (b) Low pressure space
   (c) Poor soil structure (d) All of these

16. **The anion exchange capacity (AEC) increases as soil pH.**
   (a) Remain unaffected (b) Decreases
   (c) First increases then decreases (d) First decreases then increases

17. **One acre of soil to a depth of 6 inch weight about**
   (a) 2.25×10^6 kg (b) 2×10^6 kg
   (c) 2×10^6 kg (d) None of these

18. **Who considered the father of Pedology ?**
   (a) Watson (b) Mendel
   (c) Liebig (d) V.V. Dopuchaev

19. Who considered the various properties of soils in relation to plant production ?

(a) Agronomist (b) Soil chemist
(c) Edaphologist (d) Bacteriologist

20. Vertical section of a soil is known as

(a) Profile (b) Horizons
(c) A soil (d) None of these

21. The solum is comprised of

(a) A, B and E horizons (b) A, B and C horizons
(c) A and B horizons (d) A-1 regolith

22. The organic soils commonly contain more than 50% organic matter by volume and at least _____ % by weight basis.

(a) 20% (b) 30%
(c) 10% (d) 25 %

23. Wind blown materials, comprised primarily of silt some fine sand and clay, are called

(a) Eclian (b) Loess
(c) Moraine (d) None of these

24. Passive factor of soil formation includes

(a) Climate (b) Organisms
(c) Parent material (d) Alfisols

25. Which of the following soils are good for agricultural production ?

(a) Clayey (b) Loamy
(c) Sandy (d) None of these

26. Crop suitable under heavy texture soil is

(a) Cotton (b) Coriander
(c) Rice (d) All of these

27. Groundnut and potato crops generally prefer

(a) Heavy soils (b) Light soils
(c) Medium soils (d) All of these

28. Commonly followed soil particles classification in India is

(a) British Standards Institutions
(b) International Society of Soil Science
(c) United States Department of Agriculture
(d) United States Public Roads Administration

29. Which of the following mineral is exist in clay particles ?

(a) Kaolinite (b) Illite
(c) Montmorillonite (d) All of these

30. The basic property of soil is

(a) Porosity (b) Texture
(c) Structure (d) Water holding capacity

31. Generally roots occupy ____ of soil mass

(a) 1/6 (b) 1/3
(c) 1/2 (d) 1/10

32. Good structure for crop production is

(a) Granular (b) Crumbly
(c) Both (a) and (b) (d) Blocky

33. Tillage improves

(a) Aeration in soil (b) Water holding capacity
(c) Hydraulic conductivity (d) All the above

34. Which crop requires fine seed bed ?

(a) Finger millet (b) Onion
(c) Lucerne (d) All of these

35. Correct moisture condition for ploughing is

(a) Ball has to be formed by pressing in hand
(b) Ball has to be disintegrated when it dropped from height of 1 metre
(c) Both (a) and (b)
(d) None of these

36. Puddled soil has _______ higher water use efficiency than non-puddled soil

(a) 2.5% (b) 20%
(c) 15% (d) 8%

37. Conservation tillage is also known as

(a) Stubble mulch tillage (b) Eco-fallow
(c) Reduced tillage (d) All of these

38. Sod seeding is common practice in

(a) USA (b) Australia
(c) India (d) China

39. Most important tillage implement used in India is

(a) Country plough (b) Mould board plough
(c) Victory plough (d) Harrows

40. Junior hoe is used for

(a) Inter-cultivation in wide spaced crops
(b) Inter-cultivation in close spaced crops
(c) Both (a) and (b)
(d) None of these

41. Cutting, creeping and spreading grasses can be done by

(a) Country plough (b) Disc plough
(c) Both (a) and (b) (d) None of these

42. Summer plough improves

(a) Soil porosity (b) Soil moisture
(c) Soil structure (d) Conoidal property

43. Good tilth has ____ mm size of aggregates

(a) 0-2 (b) 0-3
(c) 0-4 (d) 0-5

44. Plough used for drainage purpose is

(a) Mould board plough (b) Victory plough
(c) Mole plough (d) Chisel plough

45. Wooden plough serves the purpose of

(a) Opening up soil (b) Covering manures
(c) Removal of weeds (d) All of these

46. Increased CO_2 concentration due to poor soil aeration will cause

(a) Adsorption of nutrients (b) Reduce water uptake capacity
(c) Reduce root respiration capacity (d) All the above

47. Cultivation with implements to control weeds before the crop comes up is known as

(a) Blind tillage (b) After cultivation
(c) Secondary tillage (d) None of these

48. The word tillage is derived from

(a) French (b) Latin
(c) English (d) Anglo-Saxon

49. Operation carried out from the harvest of the crop to before sowing is known as

(a) Preparatory tillage (b) Deep tillage
(c) Secondary tillage (d) None of these

50. ________ % depletion of available moisture is optimum for ploughing

(a) 0-25 (b) 25-50
(c) 50-75 (d) 60-100

51. Physical property of soil depends on ___ of soil property

(a) Size (b) pH

(c) Class (d) Moisture

52. Element responsible for binding primary particle is

(a) Ca (b) Na

(c) K (d) None of these

53. During decomposition of organic matter is the major organic acid produced

(a) Fulvic acid (b) Humic acid

(c) Polysaccharides (d) None of these

54. Specific area of clay particle is _____ times higher than sandy particle

(a) 100 (b) 500

(c) 1000 (d) 10000

55. Degree of distinctness and durability of individual peds are ____ of soil structure

(a) Type (b) Class

(c) Grade (d) None of these

56. Common criterion generally used to determine CO_2 concentration in soil pores space is

(a) O_2 content in soil (b) CO_2 content in soil

(c) Oxygen diffusion rate (d) None of these

57. Which of the nutrient concentration increases with the increase in soil moisture (i.e. under flooding) ?

(a) P (b) Fe

(c) Mn (d) All of these

58. Under water logged condition, the nutrient shows deficiency symptom in crops is

(a) Zinc (b) Copper

(c) Both (a) and (b) (d) Iron

59. Under anaerobic soil condition, the toxic acid produced is

(a) Lactic acid (b) Butyric acid

(c) Citric acid (d) All of these

60. Which of the following mineral has the lowest specific heat ?

(a) Quartz (b) Illite

(c) Kaolinite (d) Calcite

61. Who is the father of tillage ?

(a) Aurthur Young (b) Liebig

(c) Jethro Tull (d) None of these

62. Mechanical manipulation of soil with tools and implements is known as

(a) Tillage (b) Tilth
(c) Both (a) and (b) (d) None of these

63. Physical condition of soil resulted due to tillage is called

(a) Texture (b) Structure
(c) Tilth (d) None of these

64. What is the important property of soil ?

(a) Size distribution of aggregates (b) Friability of soil
(c) Both (a) and (b) (d) None of these

65. Soil aggregate size must be ______ diameter for good crop under irrigated conditions

(a) 1-3 mm (b) 2-3 mm
(c) 3-4 mm (d) >5 mm

66. Tillage operations share about ____ percentage of total cost of cultivation

(a) 15 (b) 30
(c) 40 (d) 60

67. Critical level of oxygen for better root growth is

(a) 1-2% (b) 4-6%
(c) 7-10% (d) >10%

68. Deep tillage is generally preferred under

(a) Long duration, deep rooted crops (b) Root crops
(c) Dryland conditions (d) All of these

69. Which crop responds well for deep ploughing ?

(a) Sugarcane (b) Potato
(c) Cotton (d) All of these

70. Which crop prefers shallow tillage ?

(a) Millets (b) Sesame
(c) Flax (d) All of these

71. Subsoil compaction is generally broken by

(a) Chisel plough (b) Mould board plough
(c) Disc plough (d) None of these

72. Year round tillage is one where operations are done

(a) Between the seasons (b) Throughout the year
(c) Both (a) and (b) (d) None of these

73. Which is the primary tillage operation ?

(a) Deep tillage (b) Sub soil tillage
(c) Year round tillage (d) All of these

74. Which is the secondary tillage operation ?

(a) Disc harrowing (b) Blade harrowing
(c) Both (a) and (b) (d) None of these

75. Equal inter-intra row spacing can be followed for

(a) Tobacco planting (b) Tomato planting
(c) Chillies planting (d) All of these

76. Seline planting is practised in Gujarat for

(a) Cotton cultivation (b) Groundnut cultivation
(c) Both (a) and (b) (d) None of these

77. Which of the following operation is after cultivation operation ?

(a) Inter-cultivation (b) Earthing up
(c) Closing deep cracks (d) All of these

78. Primary tillage operations aim to

(a) Improve soil moisture (b) Obtain good seed-bed for sowing
(c) Both (a) and (b) (d) None of these

79. Sub soiler works up to a depth of

(a) 20 cm (b) 45 cm
(c) 50 cm (d) 65 cm

80. Peg type weeder is generally used in _______ soils

(a) Clay (b) Loamy
(c) Both (a) and (b) (d) None of these

81. Factor influencing preparatory cultivation is

(a) Crop (b) Soil and climate
(c) Type of farming (d) All the above

82. In rice fields, weeding is practised by using

(a) Star weeder (b) Japanese rotary weeder
(c) Both (a) and (b) (d) All of these

83. Concept of minimum tillage was started in

(a) Africa (b) USA
(c) India (d) China

84. Minimum tillage can be done by

(a) Omitting the uneconomical operations
(b) Combining operations
(c) Both (a) and (b)
(d) None of these

85. The minimum tillage operation is/are

(a) Row zone tillage
(b) Plough plant tillage
(c) Wheel tract tillage
(d) All of these

86. Secondary tillage done only in seed rows is

(a) Row zone tillage
(b) Plough plant tillage
(c) Wheel tract tillage
(d) None of these

87. Drainage is ____ times slower in puddled soils than granulated soils

(a) 20
(b) 50
(c) 1000
(d) 500

88. What is the main objective of inter-cultivation ?

(a) Weed control
(b) Create dust mulch to reduce evaporation
(c) Facilitate aeration to the root system
(d All the above

89. Which soil order derived from volcanic ejecta and dominated by allophane

(a) Histosols
(b) Andisols
(c) Alfisols
(d) Ultisols

90. CEC of which of clay mineral is highest (120 - 150 c mol/ kg).

(a) Mica
(b) Vermiculite
(c) Montmorillonite
(d) Chlorite

91. Which of the following is not correctly matched

Soil order		**Pronunciation**
(a) Entisols	:	Recent
(b) Aridisols	:	Arid
(c) Histosols	:	Organic
(d) Vertisols	:	Pedalfor

92. Lowest category of soil classification includes

(a) Order
(b) Family
(c) Series
(d) None of these

93. A clay soil must contain at least ____ clay separate to be designated a clay

(a) 80% (b) 70%
(c) 50% (d) 35%

94. In which of the following sequence, soil structure is named ?

(a) Grade, class, type (b) Type, class, grade
(c) Grade, type, class (d) Class, type, grade

95. Aggregate formation is due to

(a) Silt clay contact and microbial activity
(b) Presence of organic compounds and irreversible soil colloids
(c) Long range and van der Waal's force acting on minerals and presence of salts
(d) All the above

96. Iron is acting as colloidal material due to

(a) Acting as flocculating agent (b) Cementing agent
(c) Both (a) and (b) (d) None of these

97. Best structure for favourable physical properties is

(a) Spherical (b) Platy
(c) Blocky (d) None of these

98. Bulk density of normal soil is

(a) 1.60 (b) 1.10
(c) 1.33 (d) 1.40

99. Which components constitute the largest volume of soil mass and plays vital role in crop growth ?

(a) Mineral (b) Organic matter
(c) Soil water (d) Soil air

100. Percentage of silt content in best soil (loamy) is

(a) 15-20 (b) 20-25
(c) 30-50 (d) 60-70

101. Which of the following soil structure warms up very quickly ?

(a) Granular (b) Crumby
(c) Both (a) and (b) (d) Columnar

102. Most controlling factor of soil temperature is

(a) Soil texture (b) Soil structure
(c) Soil moisture (d) Soil consistence

103. Structure which has more porosity is

(a) Granular (b) Crumby
(c) Both (a) and (b) (d) Platy

104. Ratio of macro and micro pores must be _________ for an ideal soil

(a) 1:1 (b) 1:2
(c) 2:1 (d) 3:1

105. In the presence of which of the following mineral in the soil, particle density is higher?

(a) Magnetite (b) Limonite
(c) Hematite (d) All of these

106. Which of the following thermal characteristics is paying role in soil temperature ?

(a) Heat capacity (b) Thermal conductivity
(c) Thermal diffusivity (d) All of these

107. Mottled colour of soil is due to the alteration of oxidation and reduction process of compounds like

(a) Iron (b) Manganese
(c) Both (a) and (b) (d) N

108. Capillary and non capillary pores must be in _____ proportion for good crop production

(a) 60:40 (b) 50:50
(c) 75:25 (d) 80:20

109. What is the objective of Tillage ?

(a) Suitable seed bed preparation (b) Weed control
(c) Improve aeration in the soil (d) All of these

110. What is the objective of puddling ?

(a) Brings ideal soil condition for rice
(b) Reduce deep percolation losses
(c) Water and weed control
(d) All of these

111. Summer tillage operation aim to

(a) Eradicating perennial weeds (b) Checking soil erosion
(c) Destroying pest and disease (d) All the above

112. Mulch tillage is also known as

(a) Fallow tillage (b) Clean tillage
(c) Both (a) and (b) (d) None of these

113. Mould board was invented by

(a) William Amos (England) (b) James Small (Scotland)
(c) Thomas Jefferson (USA) (d) All together

114. Seed drill and horse-hoe were introduced by

(a) Thomas Jefferson (b) Jethro Tull
(c) James Small (d) None of these

115. Zero tillage (no tillage) is possible where the soil has

(a) Coarse textured surface
(b) Good internal drainage
(c) Adequate quantity of residue as mulch
(d) All of these

116. Demerits of zero tillage is

(a) Low seed germination (b) Low mineralization
(c) Built up of volunteer plant (d) All the above

117. Why modern tillage is not possible in India ?

(a) Fodder requirement (b) Fragmentation of land
(c) Lack of implements (d) All of these

118. Minimum tillage felt necessary in USA during

(a) 1960 (b) 1974
(c) 1980 (d) 1998

119. Zero tillage practised in _______

(a) Areas prone to erosion
(b) High intensive areas where there is no time for field preparation
(c) Both (a) and (b)
(d) None of these

120. Under zero tillage commonly used herbicide is

(a) Paraquat (b) Diquat
(c) Both (a) and (b) (d) None of these

121. Basic requirement of zero tillage is

(a) All vegetation is killed by herbicides
(b) Row surface is cleared-off mulch
(c) Seeds placed in proper depth with good germeability
(d) All the above

122. Iron plough was invented by

(a) James Small (b) William Amos
(c) Thomas Jefferson (d) None of these

123. Iron share was first patented in 1785 by

(a) James Small (b) William Amos
(c) Robert Ransome (d) Jethro Tull

124. Mould board ploughs are made in India in

(a) 1920 (b) 1940

(c) 1950 (d) 1960

125. Minimum tillage is beneficious over zero tillage because

(a) Higher infiltration (b) Good crop stand

(c) Lesser resistance to root growth (d) All of these

126. Objective of conservation tillage is mainly to conserve

(a) Soil (b) Water (moisture)

(c) Both (a) and (b) (d) None of these

127. Animal drawn mould board plough is also known as

(a) Victory plough (b) Ridge plough

(c) Rotary plough (d) None of these

128. Disc plough can be used in

(a) Non stony lands (b) Grass lands

(c) Both (a) and (b) (d) None of these

129. Weeding operations for closed spaced crops is done by

(a) Star weeder (b) Rotary weeder

(c) Both (a) and (b) (d) None of these

130. Per cent pore space can be calculated by using of which formula

(a) $\% PS = 100-(D_p/D_b \times 100)$ (b) $\% PS = 100-(D_b/D_p \times 100)$

(c) $\% PS = 1-(D_p/D_0 \times 100)$ (d) $\% PS = (D_p/D_p \times 100)$

131. How much quantity of total incoming solar radiation is absorbed by the soil and can be used to warm it.

(a) 30% (b) 40%

(c) 15% (d) 10%

132. Magnesium in octahedral sheet of clay mineral indicates

(a) Dioctahedral (b) Trioctahedral

(c) Tetrahedral (d) Monohedral

133. 'Peat' soil are generally characterized as

(a) Black, heavy and acidic soil

(b) Red, light and acidic soil

(c) Black, light and alkaline soil

(d) Black, heavy and neutral soil

134. What is the porosity percentage if the particle density of soil is 2.6 and bulk density is 1.3 ?

(a) 35% (b) 60%
(c) 50% (d) None of these

135. "White alkali" soils are

(a) Acid soil (b) Saline soil
(c) Saline sodic soil (d) Sodic soil

136. Which microorganisms have highest biomass in soil ?

(a) Bacteria (b) Actinomycetes
(c) Fungi (d) Algae

137. Soil aeration limits plants growth when percent pores filled with water are more than

(a) 50-60% (b) 60-70%
(c) 70-80% (d) 80-90%

138. Soil colour determination chroma refers to which of the following

(a) Dominant spectral colour
(b) Lightness or darkness of the colour
(c) Purity or strength of colour
(d) None of these

139. In the soil solution the negative logarithm of H^+ ion concentration is

(a) Alkalinity (b) pH
(c) Salinity (d) None of these

140. Crystalline metamorphic rock is

(a) Coal (b) Limestone
(c) Sandstone (d) None of these

Fill in the Blanks

1. The book " Soil Clay Minerology" was written by _________ .
2. Black soils of India are locally called as __________ .
3. The pH was given by ______ .
4. Soil pH is a measure of the _________ .
5. The C:N ratio of organic matter is _______ .
6. The C:N ratio of Indian soils is ________ .
7. Soil viruses are ________ .
8. Soil horizons A, B and C (A+B+C) are collectively called as_________ .
9. The book 'Benchmark Soils of India' was written by___________ .

10. The primary objective of tillage is _______ .
11. A crop planting system that allows minimum disturbance of the soil to allow seeds to be sown while ensuring maintenance of crop residues on the surface is known as _________.
12. Best agricultural soils of the word is ________ .
13. In India, Andisols may be observed in __________ .
14. If a soil is with high content of fulvic acid, then its colour will be _______ .
15. If a soil is with high content of humic acid, then its colour will be _______ .
16. Colour of the soil samples are measured by _________ .
17. 'LIS' describes _______ country's soil information system.
18. "Soil is mixture of earth's upper most mantles of weathered rocks and organic matter" this statement given by __________ .
19. Father of Indian Soil Science is ___________ .
20. Science that deals with soil genesis and classification is called ________ .

True/False

1. Surface soils have lighter shade. **(T)**
2. "NRIS" describes India soil information system. **(T)**
3. The book 'Fundamental of Soil Physics' was written by J. S. Kanwer. **(F)**
4. Dokuchaiev is the father of soil science. **(T)**
5. Tillage is most important to crop production because it prepares a good seed bed. **(T)**
6. The definition "Soil is a dynamic natural body" was given by Joffe (1949). **(F)**
7. Clay loam soils has high porosity. **(T)**
8. Particle density is also called as true density. **(T)**
9. Soil function as Chemical and Biological filter. **(T)**
10. The maximum water holding capacity is seen in clay soil. **(T)**
11. Saltation is a type of water erosion. **(F)**
12. The most productive soil of India is Alluvial soil. **(T)**
13. Soil formed by the process of gleization is called as hydromorphic soil. **(T)**
14. Saline soils are also called as Solonchalks. **(T)**
15. Sodic soils are also called as Soloth. **(F)**

Answers Fill in the Blanks

1. Stevenson
2. Regurs
3. Sorensen
4. Hydrogen ion concentration
5. 10:1
6. 100:1
7. Obligate parasites
8. Soil regolith

9. Murthy et al.
10. Seedbed preparation
11. Conservation tillage
12. Mollisols
13. Andman and Nicobars
14. Yellow
15. Brown
16. Munsell colour chart
17 UK
18. Ruffin
19. J.W. Leather
20. Pedology

ANSWERS

1.(c)	2.(b)	3.(a)	4.(a)	5.(a)	6.(b)	7.(a)
8.(d)	9.(d)	10.(c)	11.(b)	12.(d)	13.(d)	14.(a)
15.(d)	16.(b)	17.(c)	18.(d)	19.(c)	20.(a)	21.(a)
22.(a)	23.(a)	24.(c)	25.(b)	26.(d)	27.(b)	28.(b)
29.(d)	30.(b)	31.(d)	32.(c)	33.(d)	34.(d)	35.(c)
36.(a)	37.(d)	38.(b)	39.(a)	40.(a)	41.(c)	42.(c)
43.(d)	44.(c)	45.(d)	46.(d)	47.(a)	48.(d)	49.(a)
50.(b)	51.(a)	52.(a)	53.(b)	54.(d)	55.(c)	56.(c)
57.(d)	58.(c)	59.(d)	60.(a)	61.(c)	62.(a)	63.(c)
64.(c)	65.(d)	66.(b)	67.(a)	68.(d)	69.(d)	70.(d)
71.(d)	72.(b)	73.(d)	74.(c)	75.(d)	76.(c)	77.(d)
78.(c)	79.(c)	80.(c)	81.(d)	82.(b)	83.(b)	84.(c)
85.(d)	86.(a)	87.(c)	88.(d)	89.(b)	90.(b)	91.(d)
92.(c)	93.(a)	94.(a)	95.(d)	96.(c)	97.(a)	98.(c)
99.(a)	100.(c)	101.(c)	102.(c)	103.(c)	104.(a)	105.(d)
106.(d)	107.(c)	108.(b)	109.(d)	110.(d)	111.(d)	112.(c)
113.(d)	114.(b)	115.(d)	116.(d)	117.(d)	118.(b)	119.(c)
120.(c)	121.(d)	122.(a)	123.(c)	124.(a)	125.(d)	126.(c)
127.(a)	128.(c)	129.(c)	130.(b)	131.(d)	132.(b)	133.(a)
134.(c)	135.(b)	136.(c)	137.(a)	138.(c)	139.(b)	140.(a)

4

Soil and Nutrients Management

1. **Among major plant nutrients the deficiency of the following nutrients in much common in soils of India**
   (a) Nitrogen (b) Potassium
   (c) Both (a) and (b) (d) None of these
2. **The alluvial soils of India, deficiency of the following nutrients is most wide spread :**
   (a) Iron (b) Zinc
   (c) Copper (d) Manganese
3. **Red soils to India are poor in**
   (a) Nitrogen (b) Phosphorus
   (c) Both (a) and (b) (d) None of these
4. **Good 'indicator plant' for nitrogen deficiency is**
   (a) Rice (b) Maize
   (c) Sunflower (d) Wheat
5. **Indicator plant for monitoring K-deficiency is**
   (a) Pea (b) Sunflower
   (c) Potato (d) Mustard
6. **Organic colloid present in soils is**
   (a) Clay (b) Mineral
   (c) Silt (d) Humus
7. **Residual acidity of ammonium sulphate is equivalent to**
   (a) 80 kg $CaCO_3$ 100 kg^{-1} (b) 70 kg $CaCO_3$ 100 kg^{-1}
   (c) 50 kg $CaCO_3$ 100 kg^{-1} (d) 40 kg $CaCO_3$ 100 kg^{-1}

8. **The size of coarse sand particle is**
(a) 2.0-0.2 mm (b) 0.2-0.02 mm
(c) 2.0-0.02 mm (d) More than 2 mm

9. **Which plant hormone induces uniform ripening in fruit ?**
(a) NAA (b) GA_3
(c) Ethephon (d) IAA

10. **Safe limit of Residual Sodium Carbonate (RSC) in irrigation water is**
(a) Less than 1.25 (b) 1.25-2.50
(c) More than 2.50 (d) Less than 3.50

11. **The colour of the powder of mineral is known as**
(a) Tenacity (b) Streak
(c) Translucency (d) Opacity

12. **Boron content in borax is**
(a) 12% (b) 10%
(c) 8% (d) 11%

13. **Chemical composition of quartz is**
(a) $BaSO_4$ (b) CaO
(c) $MgSO_4$ (d) SiO_2

14. **Ricinine is extracted from leaves and seeds of**
(a) Castor (b) Sunflower
(c) Safflower (d) Karanj

15. **Which are the derivatives of adenine ?**
(a) Gibberellins (b) Auxin
(c) Ethylene (d) Cytokinin

16. **Neutral normal ammonium acetate is used for extraction of ______ .**
(a) Potassium (b) Phosphorus
(c) Calcium (d) Sulphur

17. **The best soil structure for plant growth is _______ .**
(a) Granular & crumb (b) Platy & round
(c) Cylindrical & spherical (d) Angular & platy

18. **Which soil holds more water at equivalent moisture tension ?**
(a) Clayey soil (b) Sandy soil
(c) Loamy soil (d) Sandy loam soil

19. **Which water in soil is tightly bound with soil ?**
(a) Hygroscopic water (b) Capillary water
(c) Gravitational water (d) Available water

20. Jenny's equation of soil formation is _____ .

(a) S = f (u, b, r, p, t _________)
(b) S = f (cl, m, r, p, t _________)
(c) S = f (cl, b, r, p, t _________)
(d) None of the above

21. Which law is used in soil texture determination ?

(a) Fick's law (b) Stoke's law
(c) Darcy's law (d) Van der - Waal's law

22. Which clay mineral has swell-shrink property ?

(a) Illite (b) Vermiculite
(c) Smectite (d) Kaolinite

23. Which method of composting requires turnings ?

(a) Pit method (b) Heap method
(c) Trench method (d) Banglore method

24. Which is the synthetic chelated compound ?

(a) NAA (b) ICA
(c) EDTA (d) ABA

25. Which mineral is having plate like structure ?

(a) Hornblends (b) Apetite
(c) Biotite (d) Tourmaline

26. Relative proportion of soil separates is called _______ .

(a) Arrangement (b) Aggregation
(c) Ratio of separates (d) Texture

27. Systematic examination, description, classification and mapping of soil is called _____ .

(a) Soil genesis (b) Soil taxonomy
(c) Soil profile (d) Soil survey

28. What is the percentage of exchangeable sodium in sodic soil ?

(a) < 15 (b) > 15
(c) 10-15 (d) < 10

29. Addition of the following material makes it possible to take good crop in sodic soil

(a) FYM (b) Green manure
(c) Gypsum (d) All of these

30. Alluvial soils are found mostly in

(a) Kerela, Karnataka and Maharashtra
(b) A.P, M.P., and Assam
(c) U.P., Punjab, Haryana, Bihar and W.B.
(d) None of these

31. Alluvial soils are rich in

(a) K_2O, P_2O_5 and Lime
(b) N
(c) Fe, Al
(d) Si, Cu

32. Ammonia is lost through volatilization in significant quantities from

(a) Alkaline soils
(b) Acidic soils
(c) Saline soils
(d) None of these

33. Ammonium sulphate is preferred for nitrogen application in groundnut because it also provides

(a) Calcium
(b) Magnesium
(c) Sulphur
(d) Copper

34. Arrangement of soil particles is referred to as

(a) Soil structure
(b) Soil texture
(c) Soil organization
(d) None of these

35. Organic matter contains about _______ per cent organic carbon

(a) 32
(b) 18
(c) 68
(d) 58

36. Oxidation of sulphur to sulphate is carried out by

(a) *Bacillus* spp.
(b) *Aspergillus* spp.
(c) *Azotobacter* spp.
(d) *Thiobacillus* spp.

37. Phosphorus fixation refers to the phenomenon of

(a) Absorption of P_2O_5 by plant
(b) Leaching of P_2O_5 from soil by percolating water
(c) Conversion of soluble phosphorus into plant unavailable forms
(d) Anion exchange with colloidal constituents of soil

38. Saline- sodic soils has the value of

(a) EC more than 4 mmhos/ cm, ESP more than 15, pH more than 8.5
(b) EC less than 4 mmhos/ cm, ESP more than 15, pH less than 8.5
(c) EC more than 4 mmhos/ cm, ESP less than 15, pH less than 8.5
(d) EC less than 4 mmhos/ cm, ESP more than 15, pH more than 8.5

39. Soils with high pH are generally deficient in

(a) Zn and Mn
(b) B and Fe
(c) Cu and Mo
(d) Ca and Mg

40. The availability of phosphorus is maximum at soil pH of

(a) 5.5-7.5 (b) 7.0-8.0
(c) 6.0-8.0 (d) None of these

41. In no tillage systems, the surface soil layers have

(a) Higher bulk density (b) Lower bulk density
(c) Higher total pore space (d) Lower water infiltration

42. Kaolinite is a ____ type clay mineral

(a) 1:1 (b) 2:1
(c) 2:2 (d) None of these

43. Mineralization of N refers to the conversion of

(a) Organic form into inorganic form
(b) Inorganic form into organic form
(c) Organic form to proteins
(d) All of the above

44. Molybdenum is a component of

(a) Chlorophyll (b) Nitrogenase enzyme
(c) Cell wall of plants (d) None of these

45. Murate of potash contains

(a) 40% K_2O (b) 36% K_2O
(c) 50% K_2O (d) None of the above

46. Nitrification in the soil takes place by

(a) Reduction process (b) Oxidation process
(c) Biological oxidation (d) Both (a) & (b)

47. Which one of the following fertilizers is least hydroscopic ?

(a) Urea (b) $(NH_4) SO_4$
(c) DAP (d) CAN

48. Which of the following statement is correct ?

(a) Blue-green algae causes plant diseases
(b) Blue-green algae fix atmospheric nitrogen in soil
(c) Blue-green algae grows best in maize crop
(d) Dry conditions are good for growth of blue-green algae

49. Organic matter present in soil is

(a) 5% (b) 10%
(c) 15% (d) 20%

50. The irrigation water of 18 to 26 SAR is classified as

(a) Low sodium water (b) High sodium water
(c) Medium sodium water (d) Very high sodium water

51. The C/N ratio of aquatic weed is

(a) 23:1 (b) 13:1
(c) 20:1 (d) 30:1

52. Which element is essential for chlorophyll synthesis ?

(a) Calcium (b) Boron
(c) Phosphorus (d) Magnesium

53. Which chemical amendment is used for improving acidic soil ?

(a) Lime (b) Gypsum
(c) Iron pyrite (d) None of these

54. The indicator plant for boron deficiency is

(a) Sugarbeet (b) Sunflower
(c) Cabbage (d) Cauliflower

55. Water soluble phosphorus content in single superphosphate is

(a) 16% (b) 32%
(c) 24% (d) 28%

56. Which is the non symbiotic nitrogen fixing organism ?

(a) Actinomycets (b) Rhizobium
(c) Cyanobacterium (d) Acetobacter

57. Bulk density of soil is

(a) Irreversible property (b) Dynamic property
(c) Reversible property (d) Static property

58. Carbon-Nitrogen ratio of normal soil is

(a) 18:1 to 20:1 (b) 10:1 to 15:1
(c) 10:1 to 12:1 (d) 15:1 to 20:1

59. Town waste and refuse is composted by

(a) Banglore method (b) Trench method
(c) Pit method (d) Heap method

60. Nitrogen content in blood meal is

(a) 10% (b) 12%
(c) 15% (d) 8%

61. Poudrette system is used to process

(a) Cow dung (b) Sheep and Goat dropping
(c) Night soil (d) Poultry droping

62. Nitrogen content in green leaves of glyricidia is

(a) 2.5% (b) 3.5%
(c) 3.0% (d) 4.0%

63. Bone meal is used as manure in

(a) Saline soil (b) Normal soil
(c) Sodic soil (d) Acidic soil

64. Residual acidity of nitrogen in fertilizer mixture is neutralised by

(a) Dolomite (b) Calcite
(c) Haematite (d) Illite

65. Cell division in plant is induced by

(a) Auxin (b) Gibberellic acid
(c) Cytokinin (d) Ethylene

66. Which is the formula for SAR ?

(a) $Na\sqrt{\frac{Ca+Mg}{2}}$ (b) $Ca\sqrt{\frac{Na+Mg}{2}}$

(c) $Mg\sqrt{\frac{Na+Ca}{2}}$ (d) $Na\sqrt{\frac{Ca+Mg}{3}}$

67. The particular size of silt as per ISSS classification is ______ .

(a) 0.02 to 0.002 mm (b) 0.2 to 0.02 mm
(c) < 0.002 mm (d) 2.0 to 0.2 mm

68. Which law state the rate of water flow in soil ?

(a) Fick's law (b) Jenny's law
(c) Stoke's law (d) Darcy's law

69. Which of the following phosphatic fertilizer is suitable for calcareous soil ?

(a) Single superphosphate (b) Dicalcium phosphate
(c) Dicalcium phosphate (d) Rock phosphate

70. Which instrument is used in aggregate size distribution analysis of soil ?

(a) Yogard's (b) Hydrometer
(c) Barometer (d) Osmometer

71. Which is the available form of sulphur in soil ?

(a) Sulphate (b) Sulphite
(c) Sulphide (d) Sulphur dioxide

72. Organic carbon content in organic matter is _____ .

(a) 48% (b) 58%
(c) 68% (d) 38%

73. Azolla is used as green manure in

(a) Sugarcane (b) Rice
(c) Proso millet (d) Finger millet

74. **Which instrument is used to determine the potassium and sodium ?**

(a) Spectrophotometer (b) Flame photometer
(c) UV-spectrophotometer (d) Absorption spectrophotometer

75. **Atomic absorption spectrophotometer is commonly used to determine**

(a) Macro nutrient (b) Micro-nutrient
(c) Secondary nutrient (d) None of these

76. **The element of same atom number but different mass number is called**

(a) Isobar (b) Isotopes
(c) Nucleon (d) Isotones

77. **Silica content of sub acid rock is**

(a) 50-55% (b) 60-65%
(c) 70-75% (d) 45-50%

78. **Argentimetric method is used to determine**

(a) Chloride (b) Sulphate
(c) Carbonate (d) Bicarbonate

79. **Which of the following types of soil are found more in India ?**

(a) Black soil (b) Red soil
(c) Laterite soil (d) Alluvial soil

80. **Which is the sulphur containing amino acid ?**

(a) Valine (b) Arginine
(c) Methionine (d) Tryptophan

81. **Astringency in fruits is due to**

(a) Tannins (b) Alkaloids
(c) Pectins (d) Isoprenoids

82. **Which soil structure is the best for plant growth ?**

(a) Granular & crumb (b) Platy & round
(c) Cylindrical & spherical (d) Angular & platy

83. **The capacity of soil being moulded in to any desired shape is known as**

(a) Cohesicity (b) Tenacity
(c) Elasticity (d) Plasticity

84. **The energy with which water is held by soil is called**

(a) pH (b) pF
(c) Rh (d) pOH

85. **Make weight material added to fertilizer is called**

(a) Conditioner (b) Filler
(c) Neutralizer (d) Starter

86. Which nitrogen inhibitor is used in nitrogenous fertilizer ?

(a) Neocil (b) Nicotine
(c) Nitrin (d) Nitrapirin

87. Gravitational water is designated as

(a) Unavailable water (b) Desirable available water
(c) Superfluous water (d) Difficulty available water

88. Which clay mineral shows swell-shrink property?

(a) Illite (b) Vermiculite
(c) Smectite (d) Kaolinite

89. Organic carbon content in organic matter is

(a) 48% (b) 58%
(c) 68% (d) 38%

90. Soil alkalinity is controlled by

(a) Calcium carbonate (b) Calcium sulphate
(c) Calcium hydroxide (d) Calcium oxide

91. Dimethoate is marked in the trade name of

(a) Schardan (b) Thiothion
(c) Cythion (d) Rogar

92. Maximum biuret per cent by weight in urea fertilizer is

(a) 1.5% (b) 2.5%
(c) 2.0% (d) 1.8%

93. Which is the carbon containing fertilizer ?

(a) Ammonical fertilizer (b) Nitrate fertilizer
(c) Amide fertilizer (d) Ammonical + Nitrate fertilizer

94. Which is the phosphate containing mineral?

(a) Illite (b) Kaolinite
(c) Apetite (d) Muscavite

95. Which is the precursor of auxin ?

(a) Carotin (b) Tripsin
(c) Zeatin (d) Tryptophan

96. The carbon nitrogen ratio of normal soil is ______ .

(a) 18:1 to 20:1 (b) 20:1 to 25:1
(c) 10:1 to 12:1 (d) 15:1 to 20:1

97. Electrical conductivity of saline soil is ______ .

(a) 2.5 dSm^{-1} (b) < 4 dSm^{-1}
(c) > 4 dSm^{-1} (d) 3.5 dSm^{-1}

98. Soil alkalinity is decreased by using _______.

(a) Gypsum (b) Lime
(c) Calcium hydroxide (d) Calcium oxide

99. Which method is used to compost town waste and refuse ?

(a) Banglore method (b) NADEP method
(c) Pit method (d) Heap method

100. Which is the amorphous clay mineral ?

(a) Kaolinite (b) Muscovite
(c) Vermiculite (d) Allophane

101. Turbidimetric method is used to determine ______.

(a) Chloride (b) Sulphate
(c) Carbonate (d) Bicarbonate

102. The physical condition of fertilizer is improved by _______.

(a) Stabilizer (b) Conditioner
(c) Mixer (d) Neutralizer

103. The size of coarse sand particle is _______.

(a) 2.0 - 0.2 mm (b) 0.2 - 0.02 mm
(c) 0.02 - 0.002 mm (d) > 2.0 mm

104. Residual acidity of ammonium sulphate is equivalent to ________.

(a) 80 kg $CaCO_3$100 kg^{-1} (b) 70 kg $CaCO_3$100 kg^{-1}
(c) 50 kg $CaCO_3$100 kg^{-1} (d) 40 kg $CaCO_3$100 kg^{-1}

105. Which is the potassium containing clay mineral ?

(a) Feldspar (b) Tourmalin
(c) Sylvinite (d) Dolomite

106. Which indicator is used in calcium + magnesium estimation ?

(a) Phenolphtaline (b) EBT
(c) Calcon (d) Mixed

107. Which method is used for determination of nitrogen from plant tissue ?

(a) Macrodistillation (b) Microdistillation
(c) Wet oxidation distillation (d) Dry ashing

108. Soil test crop response correlation approach for targeting crop yields is putforth by ______.

(a) Ramamoorthy (b) Gopalkrishnan
(c) Goswami (d) Rai

109. Which tissue testing technique identifies the nutritional factor that limits the crop production ?

(a) Rapid tissue testing (b) Tissue testing in laboratory
(c) Crop log (d) DRIS

110. Biological nitrogen fixation is

(a) Aerobic (b) Anaerobic
(c) Both (a) & (b) (d) None of these

111. Farm yard manure

(a) Balanced food for plants
(b) Recommended for salinity reclamation
(c) Tends to make soil acidic
(d) Tends to make soil alkaline

112. For managing acid soils following chemical amendment is quite essential

(a) $ZnSO_4$ (b) CuO
(c) CaO (d) $CaSO_4 \cdot 2H_2O$

113. For reclamation of alkaline soil, gypsum is mixed in the soil at the depth of

(a) 100 cm (b) 30 cm
(c) 60 cm (d) 10 cm

114. Gypsum, an amendment applied to alkali soils belongs to

(a) Acid former (b) Low solubility
(c) Soluble Ca- salt (d) None of these

115. Highly available form of phosphorus at pH 7.5 is

(a) H_2PO_4 (b) PO_3
(c) HPO_4 (d) Inositol

116. Humus colloids are composed basically of

(a) Carbon (b) Nitrogen
(c) Helium (d) Ozone

117. If the C: N ratio of crop residues is 20:30 then

(a) Mineralization may not take place
(b) Immobilization may take place
(c) Immobilization may not take place
(d) Both immobilization and mineralization take place

118. The example of amide group nitrogenous fertilizer

(a) Ammonium sulphate (b) Urea
(c) Calcium ammonium nitrate (d) None of the above

119. The fertilizers which have two or more than two major plant nutrient in their chemical composition are called ?

(a) Double fertilizer (b) Complex fertilizer
(c) Mixed fertilizer (d) Organic fertilizer

120. The highest nitrogen content among the commonly used nitrogenous fertilizers is embodied by

(a) Anhydrous ammonia (b) Urea
(c) Ammonium nitrate (d) Calcium ammonium nitrate

121. The most abundant mineral present in the earth's crust is

(a) Quartz (b) Felspar
(c) Mica (d) Hornblende

122. The number of essential elements required for the growth of higher plants is

(a) 3 (b) 20
(c) 8 (d) 16

123. Which micronutrient is applied through seed treatment ?

(a) Boron (b) Iron
(c) Molybdenum (d) Copper

124. Molecular weight of substance divided by valency is the

(a) Equivalent weight (b) Formula weight
(c) Chemical weight (d) Atomic weight

125. Which extractant is used for soil available phosphorus ?

(a) Sodium bicarbonate (b) Sodium carbonate
(c) Sodium sulphate (d) Sodium thiosulphate

126. The relative ease of movement of water within the soil is called

(a) Infiltration (b) Permeability
(c) Drainage (d) Hydraulic conductivity

127. Which mineral contains boron ?

(a) Rhodonite (b) Hematite
(c) Tourmaline (d) Quarzite

128. The soil aggregates are also called ______ .

(a) Primary soil particle (b) Tertiary soil particle
(c) Secondary soil particle (d) None of the above

129. The "A" horizon is also called as zone of ______ .

(a) Illuviation (b) Eluviation
(c) Solum (d) Bed rock

130. Redish colour of laterite soil is because of ______ .

(a) Oxides of manganese (b) Organic matter
(c) Iron oxide (d) Humus

131. The colour of the powder of mineral is known as ______ .

(a) Tenacity (b) Streak
(c) Translucency (d) Opacity

132. Which is the sulphur containing amino acid ?

(a) Valine (b) Arginine
(c) Methionine (d) Tryptophan

133. Arrangement of soil separates is referred as _______ .

(a) Aggregation (b) Texture
(c) Structure (d) Soil arrangement

134. Which chemical is used to hasten the decomposition in pit method ?

(a) Ammonium chloride (b) Muriate of potash
(c) Diammonium phosphate (d) Ammonium phosphate

135. Which hormone induces the cell division in plant ?

(a) Auxin (b) Gibberellin
(c) Cytokinin (d) Ethylene

136. What is the biuret per cent in urea fertilizer ?

(a) 1.5% (b) 2.5%
(c) 2.0% (d) 1.8%

137. Make weight material added to fertilizer is called ________ .

(a) Conditioner (b) Filler
(c) Neutralizer (d) Starter

138. Residual acidity of nitrogen in fertilizer mixture is neutralized by ______ .

(a) Dolomite (b) Calcite
(c) Haematite (d) Illite

139. Which is the phosphorus containing mineral ?

(a) Illite (b) Kaolinite
(c) Apetite (d) Muscovite

140. Nitrogen content in diammonium phosphate is ________ .

(a) 28% (b) 18%
(c) 46% (d) 21%

141. Blue green algae fix nitrogen mainly in

(a) Rice field (b) Cereals
(c) Sugarcane (d) Legumes

142. Bone meal is well suited for

(a) Alkaline soils (b) Neutral soils
(c) Acidic soils (d) All of these

143. Buffering capacity of soil means resistance to a change in

(a) Soil microbial (b) Soil acidity
(c) Soil pH (d) Base saturation

144. Calcium is essential for

(a) Cell walled formulation (b) Meristematic activity in roots
(c) Both (a) and (b) (d) None of these

145. Cation exchange capacity is expressed as

(a) Mole/litre (b) Meq/ 100g
(c) Meq/lit (d) Per cent

146. Denitrification is more in

(a) Waterlogged soils (b) Well drained soils
(c) Heavy soils (d) Light soils

147. In a waterlogged soil the concentration of _____ is high.

(a) Ethane (b) Methane
(c) Carbon dioxide (d) Carbon monoxide

148. In acid soils, phosphorus is fixed as

(a) Iron and aluminium phosphate (b) Calcium phosphate
(c) Fluroapatite (d) Hydeoxyapatite

149. The organic matter of soils is ordinarily obtained by multiplying the organic carbon content by

(a) 1.52 (b) 1.62
(c) 1.72 (d) 1.82

150. Which nitrogenous fertilizer will you recommend for paddy crop ?

(a) Nitrate fertilizer (b) Amide fertilizer
(c) Ammonical fertilizer (d) All of the above

151. Which of the following element is most mobile in soil ?

(a) Ca (b) Mg
(c) K (d) N

152. Which of the following is a high salt tolerant crop ?

(a) Lucern (b) Moong
(c) Urd (d) Arhar

153. Which of the following is a nitrification inhibitor ?

(a) A. M. (2, aminoa-4, 6 methyl pyridine)
(b) Thio-Urea
(c) Sulphur coated urea
(d) CBC (b-chloroethyl, trimethyl ammonium chloride)

154. Which of the following is considered as biofertilizer ?

(a) Farm yard manure
(b) Compost
(c) Green manure
(d) Blue-green algae

155. Which of the following is wrong ?

(a) Iron, magnesium, copper, zinc, boron, molybdenum, chloride and cobalt are called micronutrients
(b) Nitrogen, phosphorus and potassium are primary nutrients
(c) Calcium, magnesium and sulphur are secondary nutrients
(d) Nitrogen is converted into nitrate by earthworms

156. Which is the non expanding 2:1 silicate clay mineral ?

(a) Montmorillonite
(b) Kaolinite
(c) Smecitite
(d) Illite

157. Nitrogen content in neem cake is ______ .

(a) 1.5%
(b) 2.5%
(c) 5.2%
(d) 10.5%

158. Which of the following crops is salt tolerant ?

(a) Soybean
(b) Groundnut
(c) Wheat
(d) Pigeonpea

159. Nitrogen content in anhydrous ammonia is _______ .

(a) 60%
(b) 75%
(c) 70%
(d) 82%

160. Nitrogen loss from FYM is prevented by _____ .

(a) Gypsum
(b) Lime
(c) Calcium oxide
(d) Calcium carbonate

161. Concentration of salts in irrigation water = EC (d Sm^{-1}) x ______ .

(a) 540
(b) 640
(c) 340
(d) 440

162. Neutral normal ammonium acetate is used to extract

(a) Potassium
(b) Phosphorus
(c) Sulphur
(d) Calcium

163. Trench method of preparing the FYM is advocated by

(a) Dr. T.D. Biswas (b) Dr. C.A. Black
(c) Dr. S.C. Shrivastava (d) Dr. C.N. Acharya

164. Boron content in borax is _______ .

(a) 12% (b) 10%
(c) 8% (d) 11%

165. Amide form of nitrogen is present in _________ .

(a) Ammonium sulphate (b) Diammonium phosphate
(c) Urea (d) Ammonium nitrate

166. Content of K_2O in muriate of potash is _______ .

(a) 40% (b) 60%
(c) 70% (d) 80%

167. Which plant nutrient move from soil to the roots interaction, mass flow and diffusion in equal proportion ?

(a) Zinc (b) Boron
(c) Sulphur (d) Iron

168. Which organisms are microaerobic, free fixers or in association with grasses ?

(a) Azotobacter (b) Azospirillum
(c) Rhizobium (d) Frankia

169. Major form of boron absorbed by plants is

(a) $B_4O_7^{-2}$ (b) $H_2BO_3^-$
(c) BO_3^{2-} (d) H_3BO_3

170. When Fe contents in plants falls below or equal to _____ ppm, deficiency is likely to occur

(a) 80 (b) 50
(c) 40 (d) 110

171. Fe-deficiency is mostly seen in crops growing on

(a) Alkaline/ Calcareous soils (b) Salt affected soils
(c) Acidic soils (d) Neutral soils

172. Fe is an important component of enzymes

(a) Nitrogenase (b) Nitrate reductase
(c) Nitrite reductase (d) All of these

173. Bronzing is associated with Fe levels greater than 300 ppm in the leaf blade of

(a) Wheat (b) Pea
(c) Maize (d) Rice

174. Crinkle leaf of cotton is caused by

(a) Fe-toxicity
(b) B-toxicity
(c) Mn-toxicity
(d) Zn-toxicity

175. In plants, deficiency of Zn are usually associated with concentrations of less than

(a) 8 ppm
(b) 40 ppm
(c) 20 ppm
(d) None of these

176. Downey mildew of pearlmillet may be suppressed by increased soil levels of

(a) K
(b) Cu
(c) Cl
(d) Zn

177. Intracellular deposits of which element is known as plant opals ?

(a) Sodium
(b) Silicon
(c) Nickel
(d) Vanadium

178. Stem melanosis disease occurs in wheat due to deficiency of

(a) Mn
(b) Zn
(c) Fe
(d) Cu

179. Maximum nitrification takes place at

(a) 20% O_2
(b) 10 % O_2
(c) 30% O_2
(d) 25% O_2

180. NH_4^+ fixation takes place in

(a) Vermiculite and Illite
(b) Montmorillonite and Kaolinite
(c) Chlorite and Kaolinite
(d) All of these

181. The critical value for K- deficiency in maize is

(a) 0.5%
(b) 2.0%
(c) 1.5%
(d) 1.0%

182. Sulphur content (%) in potassium sulphate is

(a) 17
(b) 40-50
(c) 70-90
(d) 200-250

183. Sulphur soil testing is determined by using

(a) $CaCl_2$
(b) $ZnSO_4$
(c) NH_4OAC
(d) $NaHCO_3$

184. Which method is applied for testing of Fe, Zn, Cu and Mn ?

(a) EDDHA
(b) HEDTA
(c) Oxalic acid
(d) DTPA

185. Which one of the following is not a Mn mineral ?

(a) Pyrolusite (b) Hausmaninte
(c) Rhodonite (d) Tenorite

186. _______ toxicity is probably the important growth limiting factor in acid soils, particularly when pH <5.0 to 5.5

(a) Zn (b) Cu
(c) Al (d) Fe

187. Which of the following soils have poor physical condition ?

(a) Saline (b) Sodic
(c) Saline-alkali (d) Alkaline

188. Sodium adsorption ratio (SAR) can be calculated by using formula

(a) $SAR = Na^{+}\sqrt{Ca^{2+} + Mg^{2+}}$

(b) $SAR = Na^{+}\sqrt{Ca^{2+}Mg^{2+}/}$

(c) $SAR = Ca^{2+} + Mg^{2+} / \sqrt{Na^{+}}$

(d) $SAR = Na^{+}\sqrt{Ca^{2+}K/2}$

189. Who suggested that the principal nourishment of plant was water

(a) Pietro de Crescenzi (b) J. B. Helmont
(c) Francis Bacon (d) Jethro Tull

190. Who gave the statement 'water was the sole nutrient of plants'

(a) Robert Boyle (b) N. C Brady
(c) J. R. Glauber (d) J.B. Helmont

191. Leaching requirement can be estimated by formula

(a) $LR = EC_{iw}/EC_{dw}$ (b) $LR = EC_{dw}/EC_{iw} \times 100$
(c) $LR = D_{iw}/D_{dw}$ (d) Both (c) and (b)

192. _______ refers to a condition in which a crop needs more of a given nutrient yet has shown no deficiency symptom

(a) Bare economic optimum (b) Luxury consumption
(c) Hidden hunger (d) Toxicity of nutrient

193. The nutrient concentration range in which added nutrient will not increase yield but can increase nutrient concentration is referred as

(a) Toxic range (b) Critical range
(c) Luxury consumption (d) None of these

194. Which elements is considered as energy currency for the plant

(a) P (b) N
(c) Mo (d) K

195. The purpling along the leaf edges in the characteristics deficiency symptom of

(a) P (b) Fe
(c) K (d) Zn

196. Which of the following element is considered to be an immobile element in the plant ?

(a) Ca (b) Mg
(c) B (d) K

197. Availability of which micro-nutrient increases with increase in soil pH ?

(a) Mo (b) B
(c) Zn (d) Fe

198. Concentration of which element is highest in soil

(a) Fe (b) O_2
(c) Cu (d) Zn

199. Laterization is a process of accumulation in soil

(a) Iron and aluminum oxides (b) Silica
(c) $CuSO_4$ (d) $CaCO_3$

200. Podzolization is a process of accumulation in soil

(a) Silica (b) Iron and aluminum oxides
(c) $CaCO_3$ (d) $CuSO_4$

201. A sequence of soils of about the same age derived from similar parent material, and occurring under similar climatic conditions but having different characteristic due to variation in relief and drainage is called

(a) Ca^+clays (b) Catena
(c) Truncated profile (d) Caliche

202. Soil microorganism are most active at

(a) 22-25° C (b) 26-28° C
(c) 30-32° C (d) 34-36° C

203. VAM is

(a) Bacteria (b) Algae
(c) Virus (d) Fungi

Fill in the blanks

1. The nutrient responsible for development of root is _______ .
2. The soil pH for rice cultivation is _______ .

3. The enzymes required for nitrogen fixation is __________ .
4. Hydroponics is growing plants in _______ .
5. Nodule formation of legume roots is reduced during deficiency of _______ .
6. The largest reservoir of nitrogen is in __________ .
7. Indicator plant for monitoring of K deficiency is ______ .
8. Good indicator plant for 'N' deficiency is ___________ .
9. The phosphorus in plant seeds is mainly stored as _______ .
10. The essentiality of potassium to higher plants was at first reported by _______ .
11. The colour of the red soils is due to _______ .
12. Nitrogen as starter solution is applied in _______ .
13. Black cotton soils of India have originated from the rocks ________ .
14. Exanthema is due to deficiency of _______ .
15. Inteternal cork of apple/browning of cauliflower is deficiency symptoms of _______ .
16. Premature leaf fall is due to deficiency of ______ .
17. Most of the plants obtain nitrogen from the soil in the form of ______ .
18. Whiptail of crucifers is due to deficiency of ________ .
19. Nitrogen fixing bacteria were discovered by _______ .
20. Stem and root tips die in deficiency of _______ .

True/False

1. N, P, K proportion in balanced fertilization is 6:3:1. **(T)**
2. Deficiency symptoms of nitrogen in plants appear at first in top young leaves. **(F)**
3. Nitrogen is an essential constituent of the chlorophyll compound in plants. **(T)**
4. Minerals are absorbed in the form of compounds. **(F)**
5. Minerals absorbed by roots move to the leaf through xylem. **(T)**
6. Zinc is essential for stomatal closing. **(F)**
7. Soil fertility is reduced by intensive agriculture. **(T)**
8. The concept of soil as a natural body was first developed by V.V. Dakuchaev. **(T)**
9. Nitrogen fixation is a exothermic process. **(F)**
10. The red colour of legume nodules is due to the presence of Leg haemoglobin. **(T)**
11. The critical limit of boron in soil is 2.0 ppm. **(T)**

12. New alluvial plains are known as khadar. **(T)**
13. Drought resistance is increased by N element. **(F)**
14. Soil salinity is measured by conductivity meter. **(T)**
15. Top soil is darker and contains more organic matter. **(T)**

Answers Fill in the Blanks

1. Phosphorus
2. 5.5 – 6.5
3. Nitrogenase
4. Liquid culture medium
5. Sulphur and Boron
6. Air
7. Potato
8. Rice
9. Phytin
10. Sprengel
11. Wide diffusion of iron
12. Vegetable Crops
13. Deccan trap
14. Copper
15. Zinc
16. Nitrogen
17. Nitrogen
18. Mo
19. Winogradsky
20. Calcium

ANSWERS

1.(a)	2.(b)	3.(c)	4.(b)	5.(c)	6.(d)	7.(a)
8.(a)	9.(c)	10.(a)	11.(b)	12.(d)	13.(d)	14.(a)
15.(d)	16.(a)	17.(a)	18.(a)	19.(a)	20.(c)	21.(b)
22.(c)	23.(b)	24.(c)	25.(c)	26.(d)	27.(d)	28.(b)
29.(c)	30.(c)	31.(a)	32.(a)	33.(c)	34.(a)	35.(d)
36.(d)	37.(c)	38.(b)	39.(a)	40.(a)	41.(a)	42.(a)
43.(a)	44.(b)	45.(d)	46.(c)	47.(d)	48.(b)	49.(a)
50.(b)	51.(b)	52.(d)	53.(a)	54.(b)	55.(a)	56.(c)
57.(b)	58.(c)	59.(a)	60.(b)	61.(c)	62.(c)	63.(d)
64.(a)	65.(c)	66.(c)	67.(a)	68.(d)	69.(c)	70.(a)
71.(a)	72.(b)	73.(b)	74.(b)	75.(b)	76.(b)	77.(b)
78.(a)	79.(d)	80.(c)	81.(a)	82.(a)	83.(d)	84.(b)
85.(b)	86.(d)	87.(c)	88.(c)	89.(b)	90.(b)	91.(d)
92.(a)	93.(c)	94.(c)	95.(d)	96.(c)	97.(c)	98.(a)
99.(a)	100.(d)	101.(b)	102.(b)	103.(a)	104.(a)	105.(c)
106.(b)	107.(b)	108.(a)	109.(d)	110.(a)	111.(b)	112.(c)
113.(d)	114.(c)	115.(c)	116.(a)	117.(a)	118.(b)	119.(b)
120.(b)	121.(c)	122.(b)	123.(c)	124.(a)	125.(a)	126.(b)
127.(c)	128.(c)	129.(b)	130.(c)	131.(b)	132.(c)	133.(c)
134.(d)	135.(c)	136.(a)	137.(b)	138.(a)	139.(c)	140.(b)
141.(a)	142.(c)	143.(c)	144.(c)	145.(b)	146.(a)	147.(b)
148.(a)	149.(c)	150.(b)	151.(d)	152.(a)	153.(a)	154.(d)
155.(d)	156.(d)	156.(c)	157.(c)	158.(c)	159.(d)	160.(a)
161.(b)	162.(a)	163.(d)	164.(d)	165.(c)	166.(b)	167.(a)
168.(b)	169.(d)	170.(b)	171.(a)	172.(d)	173.(d)	174.(c)
175.(c)	176.(c)	177.(b)	178.(d)	179.(a)	180.(d)	181.(d)
182.(a)	183.(a)	184.(d)	185(d)	186.(c)	187.(b)	188.(b)
189.(c)	190.(d)	191.(a)	192.(c)	193.(c)	194.(a)	195.(a)
196.(a)	197.(a)	198.(b)	199.(a)	200.(a)	201.(b)	202.(d)
203.(d)						

5

Irrigation Water Management

1. **Which of the following criteria of irrigation scheduling is more acceptable to farmers ?**

   (a) Soil moisture tension (b) Pan evaporation
   (c) Critical growth stage (d) Water potential

2. **Knot is the unit of measurement of _______ .**

   (a) Humidity (b) Temperature
   (c) Cloud (d) Wind speed

3. **Higher water use efficiency was observed in _______ irrigation method.**

   (a) Sprinkler (b) Flood
   (c) Furrow (d) Drip

4. **Water risen in a capillary tube is characterized by a**

   (a) +ve pressure potential (b) -ve pressure potential
   (c) Zero pressure potential (d) None of these

5. **Free water is characterized by a**

   (a) +ve pressure potential (b) -ve pressure potential
   (c) Zero pressure potential (d) None of these

6. **The pressure due to solutes is known as**

   (a) Osmotic pressure (b) Imbibitional pressure
   (c) Suction pressure (d) Vapour pressure

7. **Diffusion pressure deficit is expressed as**

   (a) DPD = OP-TP-IP (b) DPD = OP+TP
   (c) DPD = OP-TP+IP (d) None of these

8. **The potential is expressed in as**

   (a) Energy/ unit mass (b) Energy/ unit volume
   (c) Energy/ unit weight (d) All of these

9. **Soil moisture tension is expressed as**
(a) Atmospheres (b) Bars
(c) Psi (d) J kg^{-1}

10. **Tensiometer depicts soil moisture**
(a) Pressure (b) Suction
(c) Tension (d) None of these

11. **The ratio of actual water content to water content at saturation is known as ______.**
(a) Actual water content (b) Relative water content
(c) Water content at saturation (d) None of these

12. **Actual water content can be obtained by**
(a) (FW-DW/ FW) x 100 (b) (TW-FW/ TW-DW) x 100
(c) (FW-DW/ TW-DW) x 100 (d) (FW-DW/TW-DW) x 200

13. **Relative water content can be obtained by**
(a) (FW-DW/ FW) x 100 (b) (TW-FW/ TW-DW) x 100
(c) (FW-DW/ TW-DW) x 100 (d) (FW-DW/FW) x 200

14. **Water content at saturation can be obtained by**
(a) (FW-DW/ FW) x 100 (b) (TW-FW/ TW-DW) x 100
(c) (FW-DW/ TW-DW) x 100 (d) None of these

15. **Excellent device for measuring small flow of water in channel is**
(a) Parshall flume (b) V notch
(c) Co-ordinate method (d) None of these

16. **Method of irrigation having least labour requirement**
(a) Flooding (b) Check basin
(c) Basin (d) Drip irrigation

17. **Flow of water in channel using V notch is computed by formula**
(a) $Q = 0.0138H^{25}$ (b) $Q = 0.0138H^{5}$
(c) $Q = 0.0138H^{2}$ (d) None of these

18. **Automated method of surface irrigation is**
(a) Flooding (b) Check basin
(c) Basin (d) Cablegation

19. **Trickle irrigation is used synonymously with**
(a) Sprinkler irrigation (b) Drip irrigation
(c) Basin (d) Flooding

20. **The ratio between the marketable crop yield and the water used in evapotranspiration is known as**
   (a) Consumptive use efficiency (b) Water use efficiency
   (c) Field water use efficiency (d) Water storage efficiency
21. **1 M mhos/ cm equals to**
   (a) 1 ds/ m (b) 1 ppm
   (c) 1% (d) 2 mg/l
22. **1 ds/ m equals to**
   (a) 10 M mhos/ cm (b) 10 ppm
   (c) 690 mg/l (d) 710 ppm
23. **Fresh water will have salt content**
   (a) < 500 mg/l (b) 1500-5000 mg/l
   (c) 2,50,000 mg/l (d) 1000-2000 mg/l
24. **Bittern water will have salt content**
   (a) < 500 mg/l (b) 1500-5000 mg/l
   (c) 3,50,000 mg/l (d) None of these
25. **Brackish water will have salt content**
   (a) < 500 mg/l (b) 1500-5000 mg/l
   (c) 2,50,000 mg/l (d) None of these
26. **The quantity of flow of water equal to a flow which will cover one acre to a depth of one inch is called**
   (a) Acre foot of water (b) Acre inch
   (c) Acre yards (d) Acre square foot of water
27. **Absorption of water and other substances against the concentration gradient is called**
   (a) Active absorption (b) Physisorption
   (c) Passive absorption (d) Chemosorption
28. **________ involves expenditure of energy in contrast to imbibition and osmosis**
   (a) Active absorption (b) Physisorption
   (c) Passive absorption (d) Chemosorption
29. **The rate of evapo-transpiration equal to all smaller than potential evapo-transpiration as affected by the level of the available soil water, salinity, field size, or other causes is termed as mm/day.**
   (a) Actual crop evapo-transpiration (b) PET
   (c) ET (d) None of these

30. **The rate of actual evapo-transpiration is expressed in terms of as**
(a) Cm/hr (b) Mm/day
(c) M/day (d) ppm

31. **Sufficient soil moisture in _______ stage is essential for optimum productivity of black gram**
(a) Flowering and pod development stage
(b) Branching and pod development stage
(c) Branching and flowering stage
(d) At the time of sowing stage

32. **The plaster of paris blocks are sensitive for tensions between _______ .**
(a) -2 bars (b) -1 to -15 bars
(c) -8 to -10 bars (d) None of these

33. **The fiberglass blocks are sensitive for tensions _______ .**
(a) -2 bars
(b) -1 to -15 bars
(c) Entire range of available soil moisture
(d) All of these

34. **If a crop is to be irrigated at 0.75 IW/CPE ratio with 60 mm depth, it should be irrigated after ______ CPE.**
(a) 90 mm (b) 75 mm
(c) 70 mm (d) 50 mm

35. **Under saline soil conditions _____ could help in obtaining better crop stand and yield**
(a) Sprinkler irrigation (b) Surface irrigation
(c) Drip irrigation (d) Furrow irrigation

36. **At soil water potential of _____ water can move only in vapour phase**
(a) -0.1 to 0.3 bars (b) -30 bars
(c) -0.2 to -0.3 bars (d) -1 to -3 bars

37. **Use of brackish water for agriculture has become more relevant due to ______ irrigation .**
(a) Sprinkler (b) Drip
(c) Furrow (d) Micro

38. **Drainage lowers underground water table so as to facilitate increased ___**
(a) Water holding capacity (b) Root zone depth
(c) Microbial activity (d) Acidity

39. The Darcy's law expression is

(a) $q = -k \times \Delta H$ (b) $q = k \times \Delta H$

(c) $q = k \times \Delta H2$ (d) $q = k - \Delta H$

40. The Darcy's law equation the negative sign denotes

(a) The flow is in the direction of the gradient

(b) The direction of flow is towards decreasing head

(c) Both (a) and (b)

(d) None of these

41. In Darcy's law the proportionality factor K is

(a) Hydraulic conductivity

(b) Pressure gradient

(c) Hydraulic pressure

(d) None of these

42. In Darcy's law Δ H denotes

(a) Hydraulic conductivity

(b) Pressure gradient

(c) Hydraulic head drop between two heads

(d) Hydraulic pressure

43. Darcy's law is valid for

(a) Only saturated flow (b) Only unsaturated flow

(c) Both (a) and (b) (d) None of these

44. Unsaturated hydraulic conductivity is also known as

(a) Capillary conductivity (b) Pressure gradient

(c) Hydraulic head drop (d) Hydraulic conductivity

45. ______ method is direct method of measuring soil water content

(a) Gravimetric (b) Tensiometer

(c) Neutron scattering (d) Spectrophotometric

46. The movement of water after rain or irrigation after 48-72 hours, the soil is at-

(a) Incipient wilting point (b) Field capacity

(c) Permanent writing point (d) None of the above

47. Criteria of irrigation scheduling is more acceptable to farmers is

(a) Soil moisture tension (b) Critical growth stages

(c) Water potential of plant (d) None of these

48. Water available to plants to the maximum extent is the one held _______.

(a) 1/3 atmosphere
(b) Pan evaporation
(c) Between field capacity and wilting point
(d) 1/2 atmospheres

49. Hygroscopic water, which occurs as thin film held tenaciously around soil particles with a tension of 31 atmospheric or more, is

(a) Readily available to plants (b) Not available to plants
(c) Harmful to plants (d) Useful to plants

50. The moisture, which is present in the form of a continuous film outside hygroscopic water around soil particles and micropore space is termed as

(a) Available moisture (b) Interspace water
(c) Capillary water (d) Hygroscopic water

51. Infiltration rate can be obtained from cumulative infiltration verses

(a) Time (b) Depth of water
(c) Amount of water (d) All of these

52. Permeability is often used synonymous to

(a) Hydraulic head (b) Hydraulic conductivity
(c) Infiltrability (d) Permiability

53. The readiness with which a porous medium transmit water is termed as

(a) Percolation (b) Infiltration
(c) Permiability (d) None of these

54. The downward movement of water through the soil profile is termed as

(a) Percolation (b) Seepage
(c) Infiltration (d) None of these

55. The percolation rate is controlled by

(a) Hydraulic conductivity (b) Seepage
(c) Permiability (d) None of these

56. The soil depth from which the crop extracts most of the water needed to meet its evapotranspiration requirements is termed as

(a) Effective root zone depth (b) Soil moisture depth
(c) Rhizosphere (d) None of these

57. Evaporation measurement can be obtained from weight changes of

(a) Lysimeter (b) Tensiometer
(c) Pan evaporimeter (d) pH meter

58. Evaporation measurements can be obtained by using water balance equation

(a) ET = Weight change + water added - percolation
(b) ET = Weight change + water added + percolation
(c) ET = Weight change - water added - percolation
(d) None of these

59. Consumptive use can be obtained by

(a) CU= E+T
(b) CU= E+T - plant metabolic needs
(c) CU= E+T + plant metabolic needs
(d) None of these

60. Irrigation requirement can be obtained by

(a) IR = WR - ER
(b) IR = WR + ER
(c) IR = ER - (WR+S)
(d) IR = WR - (ER+S)

61. In drip irrigation has been originated

(a) USA
(b) Israel
(c) India
(d) China

62. The first drip irrigation was developed by

(a) Symcha Blase
(b) Swaminathan
(c) Panmann
(d) Vavilov

63. The most common method among surface methods of irrigation

(a) Flooding
(b) Check basin
(c) Cablegation
(d) Basin

64. The ____ method of irrigation suitable for fruit crops.

(a) Flooding
(b) Check basin
(c) Basin
(d) Drip irrigation

65. Corrugation is a type of ______ .

(a) Surface irrigation
(b) Sub surface irrigation
(c) Furrow irrigation
(d) Sprinkler irrigation

66. EC is expressed as

(a) DS/ m
(b) Kilograms
(c) ppm
(d) mg/ l

67. Irrigation water containing more than _____ ppm of boron is harmful to crops

(a) 4
(b) 3
(c) 1
(d) 6

68. Boron waters contains _______ of boron

(a) 3.0 ppm
(b) 11-15 ppm
(c) 4.0 ppm
(d) 5-10 ppm

69. **The difference in soil moisture content between field capacity and the soil moisture content in the root zone before application of irrigation water is known as**

(a) NIR (b) GIR
(c) WR (d) None of these

70. **Irrigation interval is a function of**

(a) Crop (b) Soil
(c) Both (a) and (b) (d) None of these

71. **Design travel is**

(a) Net depth of application
(b) Net depth of application x Peak period use rate
(c) Net depth of application / Peak period use rate
(d) None of these

72. **The number of days that can be allowed for applying one irrigation to a given design area during the peak period consumptive use of the crop is called**

(a) Irrigation requirement (b) Irrigation period
(c) Water requirement (d) None of these

73. **The method of irrigation where water is allowed to run over the soil surface and allowed to infiltrate is termed as**

(a) Drip irrigation (b) Surface irrigation
(c) Furrow irrigation (d) Sprinkler irrigation

74. **The first watering to the crop is known as**

(a) Kor watering (b) Delta
(c) Base irrigation (d) None of these

75. **The part of the base period during which first watering is given to crop is known as**

(a) Kor period (b) Kor depth
(c) Base period (d) Kor watering

76. **The first irrigation before sowing the crop for seed germination and seedling establishment is known as**

(a) Palco (b) Blind irrigation
(c) Base irrigation (d) Furrow irrigation

77. **_______ is also called as consumptive use of water**

(a) Absolute water requirement (b) Relative water requirement
(c) Gross water requirement (d) Water use efficiency

78. The quantity of water in ha-cm per crop season absorbed by the crop together with the evaporation from the crop producing land is called as

(a) Absolute water requirement (b) Relative water requirement
(c) Gross water requirement (d) Water use efficiency

79. Consumptive use of water includes

(a) The water used by evapotranspiration
(b) The water retained in plant body
(c) Both (a) and (b)
(d) None of these

80. The amount of water that would cover an acre of land to a depth of one foot assuming no seepage evaporation and runoff losses is called

(a) Acre foot of water (b) Acre inch
(c) Acre yards (d) Acre square inch

81. The ratio of number of hectare days required in mono culture to the number of hectare days used in inter-cropping system to produce identical quantities of each of the component crop is known as

(a) ATER (b) NAR
(c) CGR (d) EC

82. The largest quantity of water is in ______ .

(a) Oceans (b) Surface water
(c) Ground water (d) Atmospheric water

83. The percentage quantity of fresh water in the earth is

(a) 33% (b) 2.6%
(c) 0.03% (d) 78.3%

84. The quantity of water in the oceans is _______ percent

(a) 97 (b) 2.60
(c) 0.04 (d) 78.3

85. The average rainfall of India has been estimated to be ______ mm.

(a) 1194 (b) 1283
(c) 1094 (d) 1184

86. The ultimate irrigation potential of India has been estimated to be

(a) 139 m ha (b) 210 m ha
(c) 140 m ha (d) None of these

87. ________ river basin has highest ground water potential

(a) Ganga (b) Yamuna
(c) Gomati (d) Brahmaputra

88. **The diameter of water molecule is**
(a) 3×10^{-8} cm (b) 3×10^{-4} cm
(c) 3×10^{-1} cm (d) 3×10^{-5} cm

89. **The latent heat of vaporization of water**
(a) 540 calories/g (b) 410 calories/g
(c) 170 calories/g (d) None of these

90. **Water passes from liquid to gaseous state by absorbing _____ of water**
(a) 540 calories/g (b) 400 calories/g
(c) 205 calories/g (d) 110 calories/g

91. **The attraction of water molecules for each other is termed as**
(a) Adhesion (b) Cohesion
(c) Both (a) and (b) (d) None of these

92. **At ultimate wilting point the soil water potential ranges from**
(a) -60 bars (b) -30 to -45 bars
(c) -45 to -60 bars (d) None of these

93. **When water is held tightly as thin film around soil particles and can move only in vapour phase this condition is termed as**
(a) Ultimate water (b) Field capacity
(c) Hygroscopic water (d) PWP

94. **When plant is not dead but remains in wilted condition this condition is termed as**
(a) PWP (b) Capillary water
(c) Hygroscopic water (d) None of these

95. **When plant can not absorb any water and die eventually this condition is termed as**
(a) Ultimate wilting point (b) Field capacity
(c) Hygroscopic water (d) Capillary water

96. **The soil moisture characteristic curve is strongly affected by soil**
(a) Texture (b) Structure
(c) Profile (d) None of these

97. **The process of gradually drying a saturated soil sample by applying increasing suction and recording successive measurement of water content is termed as**
(a) Sorption (b) Desorption
(c) Hysteresis (d) Pedology

98. The process of gradually wetting a dry soil sample while reducing the suction is termed as

(a) Sorption
(b) Desorption
(c) Pedology
(d) None of these

99. Hydraulic conductivity is depicted as

(a) V = Kf
(b) f = VK
(c) K = Vf
(d) None of these

100. The air entry value to tensiometer cup is as high as _______ bar

(a) 0.80
(b) 0.70
(c) 0.85
(d) 0.95

101. The tensiometer are suitable for ____ soils

(a) Clay loam
(b) Sandy
(c) Silty loam
(d) All of these

102. The tensiometer are more accurate in _____ range of soil moisture

(a) Dry
(b) Wet
(c) Both (a) and (b)
(d) None of these

103. The tensiometer indicates

(a) Soil moisture content
(b) Soil moisture suction
(c) Soil moisture potential
(d) All of these

104. The nylon blocks are sensitive for tensions less then ________.

(a) -2 bars
(b) -1 to -15 bars
(c) -5 bars
(d) None of these

105. Better drainage helps crop growth most due to _______.

(a) Better aeration of the soil
(b) Better mineralisation of manures and fertilizers
(c) Better root growth
(d) All of these

106. Tillering in rice is enhanced by _______.

(a) Optimum moisture content of the soil
(b) Saturation of the soil
(c) Shallow depth of standing water
(d) Microbial activity

107. Unsaturated hydraulic conductivity is variable and depends upon

(a) Soil moisture content
(b) Pressure gradient
(c) Hydraulic head crop
(d) Hydraulic conductivity

108. Watershed Management practice helps to increase productivity in

(a) Irrigated areas (b) Rainfed areas
(c) Humid areas (d) All of these

109. Which one of the following is considered as rain water recycling

(a) Farm pond (b) Compartmental bunding
(c) Percolation period (d) None of these

110. Moisture moves through soil pores at a rate proportional to the

(a) Hydraulic gradient (b) Hydraulic head
(c) Soil porosity (d) Soil conductivity

111. Water that is available to the plants and readily absorbed is :

(a) Gravitational water (b) Hygroscopic water
(c) Surface water (d) Capillary water

112. Water that occurs as thin film (4 to 5 mili micron) and held with a tension of 31 atmosphere or more and not available to plant is

(a) Hygroscopic water (b) Capillary water
(c) Gravitational water (d) Rain water

113. Indirect methods of measuring soil water content

(a) Gravimetric (b) Volumetric
(c) Neutron scattering (d) Spectro photometric

114. Water requirement of a crop is related to its

(a) Evapotranspiration (b) Potential absorption coefficient
(c) Dry matter content (d) Water holding capacity

115. In most of the cases field capacity represents the water held at a tension of ______.

(a) 15 atmosphere (b) 1/3 atmosphere
(c) 1/2 atmosphere (d) 20 atmosphere

116. Major source of water used by the plants is

(a) Capillary water (b) Hygroscopic water
(c) Gravitational water (d) None of these

117. The first phase of percolation is

(a) Infiltration (b) Seepage
(c) Permiability (d) None of these

118. The infiltration can not exceed

(a) Percolation (b) Seepage
(c) Permiability (d) None of these

119. _______ can not exceed the rate at which water enters the soil

(a) Infiltration
(b) Seepage rate
(c) Permiability
(d) Percolation rate

120. Percolation occurs when the suction is less than _______ .

(a) 1/5 atm
(b) 1/10 atm
(c) 1/2 atm
(d) 1/3 atm

121. The downward and lateral movement of water in the soil is known as

(a) Infiltration
(b) Seepage
(c) Permiability
(d) None of these

122. Crop coefficient is the ratio between the

(a) Water requirement and irrigation requirement
(b) Irrigation requirement and potential evapotranspiration
(c) Evapotranspiration of crop and potential evapotranspiration
(d) None of these

123. Crop coefficient depends on which of the following

(a) Soil cover
(b) Soil moisture content
(c) Crop height
(d) All of these

124. Which of the following usually contributes to water table ?

(a) Percolation
(b) Seepage
(c) Infiltration
(d) None of these

125. Water requirement can be obtained by

(a) WR = IR-ER
(b) WR = IR + ER
(c) WR = IR + ER + S
(d) None of these

126. Flow of water in pipe is measured by

(a) Pipe orifices
(b) Water meters
(c) Co-ordinate method
(d) All of these

127. The method of irrigation exclusive for law land rice is

(a) Flooding
(b) Check basin
(c) Basin
(d) Sprinkler

128. Flow of water in channel is measured by

(a) Parshall flume
(b) V notch
(c) Both (a) and (b)
(d) None of these

129. The method of irrigation in which intermittent application of water to field surface under gravity flow is practiced

(a) Flooding
(b) Check basin
(c) Cablegation
(d) Surge

130. ______ is a form of gated pipe system

(a) Flooding (b) Check basin
(c) Surge (d) Cablegation

131. The major demerit of drip irrigation system is ______ .

(a) Flooding (b) Clodding
(c) Clogging (d) All of these

132. ______ is depicted as yield of marketable crop produced per unit of water used in evapotranspiration

(a) WUE (b) WCE
(c) Both (a) and (b) (d) None of these

133. Salt content of irrigation water is measured as ________ .

(a) EC (b) CEC
(c) ppm (d) PBS

134. The method of irrigation where water is allowed to fall on the plants and soil as simulated rainfall is called as

(a) Sprinkler irrigation (b) Drip irrigation
(c) Furrow irrigation (d) None of these

135. The method of irrigation where water is applied directly to the root zone is called as

(a) Drip irrigation (b) Surface irrigation
(c) Furrow irrigation (d) None of these

136. The amount of irrigation water inclusive of evapotranspiration, consumptive use and that necessary to regulate the salt concentration in the soil solution to the total plant volume of water delivered is called as

(a) Irrigation efficiency (b) Water conveyance efficiency
(c) Water application efficiency (d) None of these

137. The ratio between the amount of water delivered to irrigated plot to the total volume of water delivered at the source is called as

(a) Irrigation efficiency (b) Water conveyance efficiency
(c) Water application efficiency (d) None of these

138. The ratio between the water stored in the root zone to the amount of water delivered to irrigated plot is called as

(a) Irrigation efficiency (b) Water conveyance efficiency
(c) Water application efficiency (d) None of these

139. Which of the following is most sensitive to boron

(a) Apple (b) Potato
(c) Berseem (d) Sugarcane

140. Which of the following is most tolerant to boron

(a) Apple (b) Maize
(c) Berseem (d) Cotton

141. Which of the following is semi tolerant to boron

(a) Grape (b) Berseem
(c) Potato (d) Cotton

142. The actual crop area under irrigation in the gross command area is called

(a) Gross command area (b) Culturable command area
(c) Net command area (d) None of these

143. The depth of water required by a crop during its life cycle in the field is called

(a) Duty (b) Delta
(c) Base period (d) None of these

144. The entire duration of crop from irrigation for preparatory tillage to last irrigation is called

(a) Duty (b) Delta
(c) Base irrigation (d) None of these

145. The ratio between the irrigated area and the quantity of water applied to the crop is known as

(a) Duty (b) Delta
(c) Base irrigation (d) None of these

146. The total geographical area of India is

(a) 328 m ha (b) 308 m ha
(c) 318 m ha (d) None of these

147. The total amount of precipitation in India accounts for

(a) 400 m ha m (b) 328 m ha m
(c) 450 m ha m (d) None of these

148. Capillarity is due to the process of _______ .

(a) Adhesion (b) Cohesion
(c) Both (a) and (b) (d) None of these

149. When all macropores are filled with air, micropores filled with water and water movement is through capillary this water is termed as

(a) Capillary water (b) Hygroscopic water
(c) Available water (d) None of these

150. At field capacity

(a) All macropores are filled with water, micropores filled with air
(b) All macropores are filled with air, micropores filled with water
(c) All macropores and micropores are filled with air
(d) All macropores and micropores are filled with water

151. At field capacity the soil water potential ranges from

(a) -0.1 to -0.3 bars (b) -0.1 to -0.3 M pa
(c) -0.2 to -0.05 bars (d) None of these

152. Field capacity is the ______ of soil water availability to plants.

(a) Upper limit (b) Lower limit
(c) Middle limit (d) All of these

153. Movement of soil water takes in the which direction ?

(a) Increasing direction (b) Decreasing direction
(c) Both (a) and (b) (d) None of these

154. At PWP the soil water potential ranges from

(a) -15 bars (b) -15 M pa
(c) -0.1 to -0.3 bars (d) -1 to -3 bars

155. At soil water potential of ________ water can move only in vapour phase

(a) -0.1 to -0.3 M pa (b) -3 M pa
(c) -0.01 to -0.03 M pa (d) None of these

156. Water movement in the soil comprises of

(a) Infiltration (b) Redistribution
(c) Withdrawal (d) All of these

157. The water available to the plants is the capillary water between

(a) PWP & Hygroscopic water
(b) Field capacity and PWP
(c) Field capacity and Hygroscopic water
(d) None of these

158. The available water holding capacity can be determined by

(a) FC-PWP (b) Field capacity- hygroscopic water
(c) PWP-hygroscopic water (d) None of these

159. ______ soil has lowest quantity of available water

(a) Sandy loam (b) Silty loam
(c) Clay loam (d) Both (a) and (b)

160. ______ soil has highest quantity of available water

(a) Clay (b) Silty loam
(c) Clay loam (d) Sandy loam

161. _______ soil texture has greater quantity of available water

(a) Fine (b) Coarse
(c) Medium (d) All of these

162. The difference in soil moisture content at a given suction pressure during the process of desorption and sorption is known as

(a) Sorption (b) Desorption
(c) Hysteresis (d) Osmosis

163. The soil moisture content at a given suction pressure during the process of hysteresis is greater for

(a) Sorption (b) Desorption
(c) Osmosis (d) None of these

164. Neutron moisture meter is suitable for ______ moisture content in the field

(a) Surface (b) Profile
(c) Texture (d) Structure

165. Neutron moisture meter is not suitable for _____ moisture content in the field

(a) Surface (b) Profile
(c) Texture (d) Structure

166. Capillary water will not be available to the plants if it is held at a tension of

(a) Less than 15 atmospheres (b) 20 atmospheres
(c) More than 15 atmospheres (d) 1/2 atmosphere

167. The moisture content at which the wilting is complete and the plant die is called the

(a) Permanent wilting (b) Ultimate wilting point
(c) Wilting range (d) None of these

168. Transpiration is a process in which

(a) Evaporation (b) Water loss from soil
(c) Water loss from plants (d) None of these

169. The soil moisture content at which the plants remain wilted day and night or in a humid chamber, the moisture content is called

(a) Permanent wilting point (b) Incipient wilting point
(c) Temporary wilting point (d) Wilting coefficient

170. The phenomenon of water entry into the soil through soil surface and vertically downward process is known as

(a) Infiltration (b) Percolation
(c) Seepage (d) Evaporation

171. The water movement as vapour through soil profile occurs by the process known as

(a) Cohesion (b) Seepage
(c) Diffusion (d) Percolation

172. Vapour movement takes from warm to _____ part of the soil

(a) Wet (b) Dry
(c) Cold (d) All of these

173. Water absorbed by the plant is used for photosynthesis is about _______ .

(a) 1% (b) 90%
(c) 110% (d) 85%

174. Water absorbed by the plant escapes as vapour in the process of transpiration is about ______ .

(a) 10% (b) 99%
(c) 89% (d) 1%

175. The major constituent of vegetative biomass is

(a) Water (b) Carbon
(c) Nitrogen (d) Phosphorus

176. The ratio between water stored in the root zone to the amount of water needed in the root zone prior to the irrigation is called as

(a) Irrigation efficiency (b) Water conveyance efficiency
(c) Water application efficiency (d) Water storage efficiency

177. The ratio between the consumptive use of water by the crop to the water depleted from the root zone is called as

(a) Consumptive use efficiency (b) Water use efficiency
(c) Field water use efficiency (d) Water storage efficiency

178. The depth of water applied in the first watering to crop is called as

(a) Delta (b) Kor depth
(c) Kor watering (d) None of these

179. The soil water in terms of reducing availability is in the sequence of

(a) FC, PWP, hygroscopic water, ultimate wilting point
(b) Field capacity, hygroscopic water, PWP, ultimate wilting point
(c) PWP, hygroscopic water, ultimate wilting point, PWP
(d) None of these

180. Water potential is expressed as

(a) Volume (b) Mass
(c) Weight (d) All of these

181. 1 psi equals to

(a) 6.9 kpa (b) 2.31 feet head
(c) Both (a) and (b) (d) None of these

182. 1 k Pa equals to

(a) 1 J kg^{-1} (b) 0.99 bar
(c) 14.5 psi (d) 100 psi

183. 1 M Pa equals to

(a) 80 bar (b) 10 bar
(c) 100 psi (d) 95 psi

184. Osmotic potential is due to presence of

(a) Solutes (b) Solvents
(c) Both (a) and (b) (d) None of these

185. When soil water is at a hydroscopic pressure greater than atmosphere, its pressure potential is considered to be

(a) Positive (b) Negative
(c) Zero (d) None of these

186. The flux passing through soil surface and flowing into the profile is called

(a) Infiltration rate (b) Percolation rate
(c) Seepage rate (d) Evaporation rate

187. The ratio between evapotranspiration of crop and potential evapotranspiration is known as

(a) Water requirement (b) Irrigation coefficient
(c) Crop coefficient (d) Wilting coefficient

188. The soil infiltration capacity depends upon

(a) Texture (b) Structure
(c) Uniformity of soil profile (d) All of these

189. The infiltration rate can be measured using

(a) Infiltrometer (b) Tensiometer
(c) Spectrophotometer (d) pH meter

190. Water containing total dissolved salts more than 1.5 m mhos are called

(a) Saline water (b) Alkaline water
(c) Normal water (d) None of these

191. Based on EC irrigation water is classified into _______ categories

(a) 2 (b) 3
(c) 4 (d) 5

192. Saline waters contains _____ as the predominant salt

(a) Sodium chloride (b) Sodium carbonate
(c) Ammonium chloride (d) None of these

193. Gross irrigation requirement is

(a) NIR x Efficiency of irrigation system
(b) NIR / Efficiency of irrigation system
(c) NIR
(d) None of these

194. The negative potential is known as

(a) Suction Potential (b) Pressure Potential
(c) Both (a) and (b) (d) Matric Potential

195. The attraction of water molecules for solid surface is termed as

(a) Adhesion (b) Cohesion
(c) Surface tension (d) None of these

196. National water policy adopted in the year ______.

(a) 1980 (b) 1981
(c) 1975 (d) 1992

197. The melting point of water is

(a) 0° C (b) 273° K
(c) -273° K (d) None of these

198. Evaporation is measured in

(a) ppm/ day (b) kg/ day
(c) mm/ day (d) Cm/ day

199. The water retention capacity is greater in

(a) Black soil (b) Sand
(c) Red soil (d) Laterite

200. The grand period of rainfall in India is

(a) North-east monsoon (b) South-east monsoon
(c) South-west monsoon (d) None of these

Fill in the blanks

1. National water policy was adopted in the year ________ .
2. Average world wide irrigation efficiency is _________ % .
3. Check basin method of irrigation is used where the land slope is ________ .

4. Best method of irrigation for potato crop is ________ .
5. Ground water is considered good if the EC (ds/m) is ______ .
6. Safe irrigation water should contain boron concentration ________ .
7. For higher productivity, the irrigation frequency should be ________ .
8. ________ growth stage of maize is most sensitive to water need.
9. IW/CPE ratio as a scheduling irrigation technique was suggested by ________ .
10. ________ irrigation system gives the highest irrigation efficiency.
11. Tensiometer was first used for scheduling irrigation by ________ .
12. For higher productivity, the irrigation frequency should be ________ .
13. Gypsum block was developed by _________ .
14. Gypsum block measures the soil water tension in the range of ______ .
15. Pressure plate apparatus was developed by ________ .
16. Transpiration is maximum at _______ .
17. Discharge of water from trapezoidal weir is equal to ________ .
18. Water discharge from V notch will be equal to _________ .
19. The amount of water consumptively used by a crop during the entire growing season or crop period is called as ________ .
20. Irrigation requirement is equal to ________________ .

True/False

1. Irrigation increase the protein and decrease the fat contents of the herbage pasture. **(T)**
2. Runoff is defined as the protein of rainfall that makes the way towards outlet. **(T)**
3. In furrow method of irrigation only 1/4th of the furrow is irrigated. **(F)**
4. Flooding method of irrigation is used when field is leveled and water is free. **(T)**
5. Darcy's law is related to water potential. **(F)**
6. Darcy's law states that velocity of water flow is directly proportional to the difference of hydraulic heads and inversely proportional to the flow length. **(T)**
7. Transpiration ratio refers to the volume of water transpired by a plant to produce a unit quantity of dry matter. **(T)**
8. Unit of transpiration ratio is kg. **(F)**
9. Cell division and elongation shall be affected first under water stress conditions. **(T)**
10. During water stress proline is accumulated in plant body. **(T)**
11. Best method of irrigation for potato crop is border. **(T)**
12. pF is equal to pH. **(F)**

13. If rainfall is P and potential evapotranspiration is PET, the moisture deficit index (MDI) will be P-PET/PET x 100. **(T)**
14. Total utilizable water resources of India are 400 BCM. **(F)**
15. Total surface water availability in the country is 1953 BCM. **(T)**

Answers Fill in the Blanks

1. 1981
2. <40%
3. <1%
4. Border
5. <2
6. 5 ppm
7. High
8. Seedling
9. Parihar and Coworkers
10. Trickle method
11. Richards
12. High
13. Bouyoucos and Milk
14. 1–20 Atm
15. Richards
16. Afternoon at 2.0 PM
17. $138 \times 10^{-4} H^{5/2}$
18. $186 \times 10^{-4} LH^{3/2}$
19. Seasonal Consumptive use
20. Water requirement – (Effective rainfall Soil Water Contribution + Ground Water Contribution

ANSWERS

1.(c)	2.(d)	3.(d)	4.(b)	5.(c)	6.(a)	7.(c)
8.(d)	9.(a)	10.(b)	11.(b)	12.(a)	13.(b)	14.(c)
15.(b)	16.(a)	17.(a)	18.(d)	19.(b)	20.(b)	21.(a)
22.(c)	23.(a)	24.(c)	25.(b)	26.(b)	27.(a)	28.(a)
29.(a)	30.(a)	31.(a)	32.(c)	33.(c)	34.(b)	35.(b)
36.(b)	37.(d)	38.(b)	39.(a)	40.(c)	41.(a)	42.(c)
43.(c)	44.(a)	45.(a)	46.(b)	47.(b)	48.(c)	49.(b)
50.(c)	51.(a)	52.(b)	53.(c)	54.(a)	55.(a)	56.(a)
57.(a)	58.(a)	59.(c)	60.(d)	61.(b)	62.(a)	63.(b)
64.(c)	65.(c)	66.(a)	67.(b)	68.(d)	69.(a)	70.(c)
71.(c)	72.(b)	73.(b)	74.(a)	75.(a)	76.(a)	77.(a)
78.(a)	79.(c)	80.(a)	81.(a)	82.(a)	83.(b)	84.(a)
85.(a)	86.(a)	87.(a)	88.(a)	89.(a)	90.(a)	91.(b)
92.(d)	93.(c)	94.(a)	95.(a)	96.(a)	97.(b)	98.(a)
99.(a)	100.(c)	101.(b)	102.(b)	103.(b)	104.(a)	105.(c)
106.(b)	107.(a)	108.(b)	109.(a)	110.(a)	111.(d)	112.(a)
113.(c)	114.(a)	115.(b)	116.(a)	117.(a)	118.(a)	119.(d)
120.(b)	121.(b)	122.(c)	123.(d)	124.(a)	125.(c)	126.(d)
127.(a)	128.(c)	129.(d)	130.(d)	131.(c)	132.(a)	133.(a)
134.(a)	135.(a)	136.(a)	137.(b)	138.(c)	139.(a)	140.(c)
141.(c)	142.(b)	143.(b)	144.(c)	145.(a)	146.(a)	147.(a)
148.(c)	149.(a)	150.(b)	151.(a)	152.(a)	153.(a)	154.(a)
155.(b)	156.(d)	157.(b)	158.(a)	159.(a)	160.(a)	161.(a)
162.(c)	163.(b)	164.(b)	165.(a)	166.(c)	167.(b)	168.(c)
169.(a)	170.(a)	171.(c)	172.(c)	173.(a)	174.(b)	175.(a)
176.(d)	177.(a)	178.(b)	179.(a)	180.(d)	181.(c)	182.(a)
183.(b)	184.(a)	185(a)	186.(a)	187.(c)	188.(d)	189.(a)
190.(a)	191.(d)	192.(a)	193.(b)	194.(d)	195.(a)	196.(b)
197.(b)	198.(c)	199.(a)	200.(c)			

6

Soil and Water Conservation

1. **Soil moisture content ______**

   (a) Decrease with increase in soil moisture tension
   (b) Increase with decrease in soil moisture tension
   (c) Does not change with soil moisture tension
   (d) None of the above

2. **The most highly visible erosion is**

   (a) Gully erosion (b) Sheet erosion
   (c) Ephemeral gullies (d) Wind erosion

3. **What is no-till crop production ?**

   (a) Preparing a good seedbed for crops
   (b) Ploughing ground, but not discing
   (c) Planting the seeds in the previous year's crop residue
   (d) Discing ground prior to planting

4. **In reduction of soil erosion, which of the following is not used ?**

   (a) Mulch (b) Gravel
   (c) Previous crops residue (d) Animal waste and residues

5. **If soil is stabilized it is**

   (a) Acted on by geological soil erosion and man-made soil erosion
   (b) Never moving under any circumstances
   (c) Under the influence of construction of a new highway
   (d) Acted on only by geological soil erosion

6. **Conservation tillage is**

   (a) A crop planting system that allows minimum disturbance of the soil to allow seeds to be sown while ensuring maintenance of crop residues on the surface
   (b) A tillage operation is mainly done to prevent the weed

(c) A tillage operation is mainly done to prepare a fine seed bed
(d) None of these

7. **The important feature of soil erosion is/are**
(a) Erosion removes topsoil
(b) Reduces levels of soil organic matter
(c) It contributes to the breakdown of soil structure
(d) All of these

8. **Land suitable for cultivation in total number of classes in the group is**
(a) Ten (b) Four
(c) Nine (d) Two

9. **In land capability classification scheme, the classes V, VI and VII are suitable for which of the following**
(a) Cultivation (b) Pasture and grazing
(c) Wildlife management (d) Watershed management

10. **Which of the following formula/equation is used for the calculation of wind erosion ?**
(a) Ramser's formula (b) Chepil equation
(c) Chepil and Woodruff equation (d) None of these

11. **Rills with more than _______ depth are known as gullies**
(a) 10 cm (b) 15 cm
(c) 45 cm (d) 30 cm

12. **Which one of the following is important with regards to soil erosion by water ?**
(a) Intensity of rainfall (b) Frequency of rainfall
(c) Amount of rainfall (d) None of these

13. **An agronomist is primarily concerned with what use of soil ?**
(a) Natural beauty
(b) Acting as a filter for the hydrologic cycle
(c) Supporting crop growth
(d) Road, building and dam construction

14. **CSWCRTI is situated at**
(a) Dehradun (b) Lucknow
(c) Nagpur (d) New Delhi

15. **No-till crop production is**
(a) Preparing a good seedbed for crops
(b) Ploughing ground, but not discing
(c) Planting the seeds in the previous year's crop residue
(d) None of these

16. Which is the dominant soil moisture regime in India ?

(a) Ustic (b) Xeric
(c) Aridic (d) Udic

17. The soil conservation and management practices suitable for arid and semi-arid regions are

(a) Tied ridges and rough seed bed (b) Ridge-furrow system
(c) Both (a) and (b) (d) None of these

18. Soil erosion is a ______ stage process

(a) Four (b) Two
(c) Three (d) Eight

19. The erosive power of rainfall is mainly depends upon the factors

(a) Rainfall duration (b) Rainfall intensity
(c) Both (a) and (b) (d) None of these

20. The volume of water that runs off during an irrigation that exceeds the soil infiltrability limit is known as

(a) Gully erosion (b) Sheet erosion
(c) Splash erosion (d) Run-off

21. The particles of the running water collide with each other resulting in to further breakdown of the particles is known as

(a) Attrition (b) Abrasion
(c) Solution (d) None of these

22. What is the most common size of raindrop ?

(a) 7 mm (b) 4 mm
(c) 5 mm (d) None of these

23. The types of clay minerals will provide more stable aggregates is

(a) Non-expanding type of clay minerals
(b) Expanding type of clay minerals
(c) Both (a) and (b)
(d) None of these

24. The most important index of soil erodibility is/are

(a) Silica content (b) Sesquioxide content
(c) Silica-sesquioxide ratio (d) All of these

25. The major role of inter-cultivation tillage is

(a) Improvement of soil structure (b) Enhancement of soil aeration
(c) Weed control (d) None of these

26. Match the following

a.	Geological erosion	i.	Severe limitations
b.	Terraces	ii.	Wearing away of land's surface by water, wind and ice
c.	Soil erosion	iii.	Wearing away of earth's surface under natural environmental conditions
d.	Class IV	iv.	Control of Mechanical erosion

(a) i, ii, iii, iv (b) iii, iv, ii, i
(c) ii, iii, i, iv (d) None of these

27. Removal of thin layers (uniform) of soil from a more or less smooth slope, carried by the distributed (rather than concentrated) flow of runoff water over the soil surface is called

(a) Sheet erosion (b) Wind erosion
(c) Gully erosion (d) Rill erosion

28. Splash and sheet erosion are also known as

(a) S inter-rill erosion (b) Gully erosion
(c) Wind erosion (d) Sheet erosion

29. The scouring and transport of soil by a concentrated flow of water is known as

(a) Rill erosion (b) Sheet erosion
(c) Gully erosion (d) Wind erosion

30. Saltation occurs when particle size ranges between

(a) 0.1-0.5 mm (b) 0.5-1 mm
(c) 6-8 mm (d) 2.0-3.0 mm

31. Different types of models are

(a) Mathematical models (b) Statistical models
(c) Simulation models (d) All of these

32. The erosive power of rainfall, running water or wind is called

(a) Erosivity (b) Run-off
(c) Erodibility (d) None of these

33. Soil separates has more erodability is

(a) Sand (b) Silt
(c) Clay (d) All of these

34. Common size of raindrop is _______ mm

(a) 7 (b) 2
(c) 4 (d) 3

35. Erosion can be removed by normal tillage operation is

(a) Rill erosion
(b) Gully erosion
(c) Sheet erosion
(d) None of these

36. Agroforestry interventions with soil and moisture conservation methods can reduce the soil loss up to

(a) 70%
(b) 50%
(c) 40%
(d) Both (b) and (c)

37. The soil properties that influences erodability is/are

(a) Infiltration capacity
(b) Structural stability
(c) Both (a) and (b)
(d) None of these

38. Water erosion is a ______ .

(a) 2 way process
(b) 1 way process
(c) 3 way process
(d) 4 way process

39. The water that runs off during an irrigation that exceeds the soil's infiltrability is called

(a) Gully erosion
(b) Sheet erosion
(c) Splash erosion
(d) Run-off

40. A section of the land surface that drains its surface water excess to a specified point along a stream is called

(a) Watershed
(b) Catchment
(c) Both (a) and (b)
(d) None of these

41. Vegetative windbreaks are also known as

(a) Catchment
(b) Tree belts
(c) Watershed
(d) Shelter belts

42. Signs of water erosion are

(a) Pedestals of soil supporting pebbles and plant material
(b) Convex landscape positions
(c) Both (a) and (b)
(d) None of these

43. Causes of water erosion are

(a) Lack of protection against raindrop impact
(b) Decreased aggregate stability
(c) Intense rainfall and residue cover is at a minimum
(d) Decreased infiltration by compaction or other means

44. What causes mechanical erosion ?

(a) Removal by harvest of root crops
(b) Tillage and cultivation practices that move soil

(c) Down slope
(d) All the above

45. **Causes of wind erosion are**
(a) Exposed surface soil during critical periods of the year
(b) Occurrence of wind velocities that are sufficient to lift
(c) Individual soil particles and long, unsheltered, smooth soil surfaces
(d) All of these

46. **Biological indicator of soil erosion is**
(a) Decreased microbial biomass
(b) Lower rate of respiration
(c) Slower decomposition of plant residues
(d) All of these

47. **How can soil erosion be measured ?**
(a) Visual (b) Physical/ chemical
(c) Biological indicators (d) All the above

48. **Visual indicators of soil erosion is/are**
(a) Comparisons of aerial photographs taken over time
(b) Presence of moss and algae crusts in desert/arid soils
(c) Changes in soil horizon thickness
(d) All of these

49. **The chemical indicator of soil erosion is**
(a) Decreases in soil organic matter content
(b) Increases in calcium carbonate content at the surface
(c) Provided greater content exists in subsurface layers
(d) All of these

50. **The physical indicator of soil erosion is**
(a) Measurements of aggregate stability
(b) Increasing depth of channels and gullies
(c) Both (a) and (b)
(d) None of these

51. **Clay minerals will provide more-stable aggregates is/are**
(a) Non-expanding type of clay minerals
(b) Expanding type of clay minerals
(c) Both (a) and (b)
(d) None of these

52. **The most important index of soil erodibility is/are**
(a) Silica content (b) Sesquioxide content
(c) Silica-sesquioxide ratio (d) All of these

53. What is the major role of inter-cultivation tillage ?

(a) Improvement of soil structure
(b) Enhancement of soil pH
(c) Weed control
(d) All of these

54. Soil crusting mainly reduces

(a) Acidity
(b) Infiltration
(c) Run-off
(d) Mulching

55. "Water harvesting" is

(a) Collection of runoff water into micro or macro reservoirs and recycling the store water for garden, water for livestock and water for irrigation
(b) Conservation of rain water in soil by preventing the runoff so that rain water is better utilized in dryland areas
(c) Collection of water to supplement the canal irrigation
(d) A method of preventing percolation loss of water by applying moisture barrier substances

56. Reason of poor water holding capacity of red soil is

(a) High infiltration
(b) Shallow department
(c) Coarse texture
(d) All of these

57. The downward flow of water from the surface into the soil is known as

(a) Seepage
(b) Infiltration
(c) Permeability
(d) Hydraulic conductivity

58. Vertical downward movement of water within the soil is known as

(a) Permeability
(b) Percolation
(c) Seepage
(d) Capillary

59. Clay content is higher in which of the following soil ?

(a) Red soil
(b) Desert soil
(c) Black soil
(d) Brown soil

60. Soil erosion can be prevented by

(a) A protective cover on the soil
(b) A barrier to the erosive agent
(c) Modifying the landscape to control runoff
(d) All the above

61. Scattering of detached soil particles by the impact of rain drops is

(a) Splash erosion
(b) Sheet erosion
(c) Gully erosion
(d) Rill erosion

62. Removal by rain of a very thin layer of soil from the entire surface of large area is

(a) Gully erosion
(b) Sheet erosion
(c) Rill erosion
(d) Splash erosion

63. Sheet erosion is observed on _______ land

(a) Deeply sloping (b) Moderately sloping
(c) Slopy (d) Gently sloping

64. Muddy run-off from the field is an indication of

(a) Gully erosion (b) Sheet erosion
(c) Rill erosion (d) Splash erosion

65. Formation of small channels a few centimetres deep all over the field along the watercourse is observed in

(a) Ravines (b) Slight gully erosion
(c) Moderate gully erosion (d) None of these

66. Erosion which is most highly visible is

(a) Gully erosion (b) Wind erosion
(c) Sheet erosion (d) Rill erosion

67. The largest of these soil particles is

(a) Gravel (b) Sand
(c) Silt (d) Clay

68. Which of these is not a key ingredient of soil ?

(a) Organic matter (b) Haemoglobin
(c) Minerals (d) Water

69. Percentage of organic matter in average soil is

(a) 35% (b) 5%
(c) 20% (d) 15%

70. Erosion created by the activities of man and animals is called as

(a) Natural geological erosion (b) Geological erosion
(c) Accelerated soil erosion (d) All of these

71. Run-off and soil loss can be calculated by

(a) Ramser's formula (b) Chepil equation
(c) Chepil and woodruff equation (d) None of these

72. Device used to study the run-off and soil loss under laboratory conditions

(a) Run-off plots (b) Multislot device
(c) Standing wavefumes (d) None of these

73. The quantity of soil materials of a given size in the run-off varies as the _____ power of the speed of run-off.

(a) Third (b) Fourth
(c) Fifth (d) Sixth

74. Which of the following is/are erosion permitting crops ?

(a) Maize (b) Wheat
(c) Bajra (d) All of these

75. Which of the following is/are erosion resisting crops ?

(a) Grasses (b) Legumes
(c) Cereals (d) Both (a) and (b)

76. Which soils are more susceptible to water erosion ?

(a) Coarse textured soils (b) Medium textured soils
(c) Fine textured soils (d) None of these

77. The ratio of dispersion ratio to colloidal moisture equivalent ratio is

(a) Erosion index (b) Soil erosivity
(c) Soil detachability (d) None of these

78. In Universal soil loss equation, A = RKLSCP, K denotes

(a) Soil conservation practices (b) Soil roughness
(c) Soil erodability (d) None of these

79. Which is erosion permitting crops ?

(a) Maize (b) Wheat
(c) Bajra (d) All of these

80. Which is erosion resisting crops ?

(a) Grasses and legumes (b) Rice
(c) Maize (d) Nnone of these

81. Removal by rain of a very thin layer of soil from the entire surface of large area is known as

(a) Ravines (b) Sheet erosion
(c) Rill erosion (d) Gully erosion

82. In Universal soil loss equation, A = RKLSCP, K denotes

(a) Soil conservation practices (b) Soil roughness
(c) Soil erodability (d) None of these

83. In Universal soil loss equation, A = RKLSCP, P denotes

(a) Soil coservation practices (b) Soil roughness
(c) Soil erodability (d) None of these

84. Symbol of very severe erosion phase is

(a) e5 (b) e1
(c) e4 (d) e3

85. In the equation of Universal soil loss, A = RKLSCP, where P denotes

(a) Soil conservation practices
(b) Soil roughness
(c) Soil erodability
(d) None of these

86. Growing of erosion resisting and erosion permitting crops on alternate strips of suitable width along the contour and across the slope is

(a) Strip cropping
(b) Contour farming
(c) Contour strip cropping
(d) None of these

87. What is the minimum wind speed required to initiate the movement of most erodable soil particles ?

(a) 16 km
(b) 8 km
(c) 10 km
(d) 20 km

88. Size range of soil particles in saltation type of wind erosion is

(a) 0.05-0.5 mm
(b) 0-1.5 mm
(c) 0.5-2.5 mm
(d) None of these

89. Cultivation of erosion resisting and erosion permitting crops on alternate strips of suitable width along the contour and across the slope is called as

(a) Contour farming
(b) Contour cropping
(c) Contour strip cropping
(d) None of these

90. Contour bounding is also known as

(a) Level terraces
(b) Ridge-type terraces
(c) Absorption type terraces
(d) All of these

91. Which type of soils are more susceptible to water erosion ?

(a) Coarse textured soils
(b) Medium textured soils
(c) Fine textured soils
(d) All of these

92. To initiate the movement of most erodable soil particles the minimum required wind speed is

(a) 8 km
(b) 10 km
(c) 24 km
(d) 16 km

93. In surface creep type of wind erosion the size of soil particle ranges

(a) < 0.5 mm
(b) 0.05-0.5 mm
(c) 0.5-2.0 mm
(d) None of these

94. In suspension type of wind erosion soil particles size ranges

(a) < 0.05 mm
(b) 0.5-1.0 mm
(c) 1.5-2.0 mm
(d) None of these

95. In saltation type of wind erosion the soil particles ranges

(a) 0.4-1.0 mm
(b) 1.0-2.0 mm
(c) 0.5-1.5 mm
(d) 0.05-0.5 mm

96. The most susceptible to wind erosion soil particle diameter is

(a) 0.1 mm (b) 0.3 mm
(c) 0.04 mm (d) 0.03 mm

97. The best vegetative cover against soil and water loss at Kanpur was provided by

(a) Velvet bean (b) Cowpea
(c) Mung (d) Sunhemp

98. The best vegetative cover at Kota was provided by

(a) Velvet bean (b) Cowpea
(c) Mung bean (d) Soyabean

99. Crops which are known to provide effective cover against loss of soil nutrient and moisture is/are

(a) Cowpea, mung and urad
(b) Dhaincha, sunhemp and groundnut
(c) Velvet bean, gaur and soyabean
(d) All of these

100. Important factors that influence the erodability of soil by wind

(a) Soil texture (b) Soil structure
(c) Soil water status (d) All of these

101. The crops which help to check soil erosion and conserve soil moisture are

(a) Cover crops (b) Conservation crops
(c) Erosion crops (d) None of these

102. The cover crop helps to effectively check soil erosion and conserve soil moisture by

(a) Checking movement of run-off water
(b) Increasing absorption of rain water by soil
(c) Preventing the rain and wind from beating the soil surface
(d) All of these

103. Crops which are most effective in protecting cultivated land from erosion is/are

(a) Legumes (b) Millets
(c) Cereals (d) All of these

104. Land capability classes are subdivided in to land capability subclasses based on their

(a) Major characteristics (b) Kind of managements
(c) Kind of limitations (d) All the above

105. In Chepil and Woodruff's equation of soil loss calculation by wind erosion, E = f (ICKLV), where K denotes

(a) Soil types
(b) Rainfall frequency
(c) Soil roughness
(d) Soil pH

106. The process that lowers the current and/or potential capability of soil to produce goods or services is called

(a) Soil degradation
(b) Land degradation
(c) Mass degradation
(d) All of these

107. The word desertification was first used by

(a) Chepil and Woodruff
(b) Weischner
(c) Abreville
(d) Borlaug

108. The smallest volume that can be called 'a soil' is

(a) Pedon
(b) Pans
(c) Pod
(d) Radon

109. The process that transforms the productive land into unproductive land is known as

(a) Desertification
(b) Desertization
(c) Mass movement
(d) All of these

110. Extension of tropical desert landscapes and landforms to area where they did not occur in the recent past is

(a) Desertification
(b) Desertization
(c) Mass movement
(d) None of these

111. Which one is important with regards to soil erosion by water ?

(a) Intensity of rainfall
(b) Frequency of rainfall
(c) Time of rainfall
(d) Duration of rainfall

112. Rills with more than _____ depth are generally called as gullies

(a) 10 cm
(b) 30 cm
(c) 45 cm
(d) None of these

113. Land capability subclasses are further subdivided into land capability units according to

(a) Major characteristics
(b) Kind of managements
(c) Kind of limitations
(d) Both (a) and (b)

114. According to land capability classification scheme, the classes V, VI and VII are suitable for

(a) Watershed management
(b) Pasture and grazing
(c) Wildlife management
(d) All of these

115. Total number of classes in the group, land suitable for cultivation is

(a) Eight (b) Four
(c) Six (d) Two

116. In north-east hill regions of India, shifting cultivation is also called as

(a) Jhum cultivation (b) Jhuming cultivation
(c) Both (a) and (b) (d) None of these

117. Total number of land capability classes in the land capability classification scheme

(a) Ten (b) Five
(c) Eight (d) Six

118. Lysimeter is used to measure _______ .

(a) Infiltration (b) Vapour pressure
(c) Evaporation (d) Evapotranspiration

119. Soil strength is determined by

(a) Penetrometer (b) Dynamometer
(c) Micrometer (d) Hydrometer

120. Most effective water saving irrigation method is ______ .

(a) Furrow (b) Border
(c) Sprinkler (d) Drip

121. Use of sprinkler irrigation can bring down about ______ per cent of saving in water use.

(a) 30-35 (b) 35-40
(c) 10-15 (d) 20-25

122. What is the commonly used soil : water ratio for pH determination ?

(a) 1:2 (b) 1:2.5
(c) 1:5 (d) 1:4

123. Age of soil was described by

(a) Jenny (b) Hilgrad
(c) Myers (d) None of these

124. Soils of semi arid and sub humid regions are

(a) Podzols (b) Lateritic and laterite soils
(c) Sierozen soils (d) Chernozems

125. Sprinkler irrigation system performance is considered satisfactory when the minimum value of the uniformity coefficient is _____ .

(a) 75% (b) 90%
(c) 55% (d) 85%

126. Infiltration rate is generally lower in soils of _______ .

(a) Light texture (b) Heavy texture
(c) Medium texture (d) None of these

127. When leaching is required to remove salt from the soil profile _____ irrigation method is used

(a) Check basin (b) Sprinkler
(c) Boreler (d) Furrow

128. A ratio of mass of water to mass of soil referred as ____

(a) Gravimetric water (b) Degree of saturation
(c) Volume wetness (d) None of these

129. An equation relating the height of the water-table between drains to depth of the drains and the distance between them, the hydraulic conductivity of the soil, and the rate of recharge by percolating water is called

(a) Hooghoudt equation (b) Chepil equation
(c) Renould's equation (d) Darcy equation

130. Scattering of defeated soil particles by the impact of rain drops is

(a) Splash erosion (b) Rill erosion
(c) Sheet erosion (d) None of these

131. Sheet erosion is observed on

(a) Deeply sloping land (b) Gently sloping land
(c) Slopy land (d) None of these

132. Muddy run-off from the field is a sign of

(a) Splash erosion (b) Sheet erosion
(c) Rill erosion (d) Gully erosion

133. Formation of small channels a few centimetres deep all over the field along the watercourse is observed in

(a) Rill erosion (b) Gully erosion
(c) Sheet erosion (d) Ravines

134. In Maharashtra at Sholapur the most effective check against soil and water loss was provided by

(a) North bean (b) Horse gram
(c) Both (a) and (b) (d) Cowpea

135. Percentage of the average soil in organic matter is

(a) 46% (b) 5%
(c) 35% (d) 18%

136. Soil erosion is

(a) The transpiration of soil particles from one place to another
(b) The result of terraces and diversions
(c) Not a natural phenomenon
(d) None of these

137. Diversion terraces are not suitable for slopes more than

(a) 12% (b) 8%
(c) 6% (d) 4%

138. If slopes are more than 20% than which types of terraces are suitable

(a) Diversion (b) Retention
(c) Bench (d) None of these

139. The type of waterways is/are

(a) Grassed waterways (b) Diversion channels
(c) Terrace channels (d) All of these

140. Examples of grass used for grassed waterways is/are

(a) Anjan (b) Bermuda
(c) Napier (d) All of these

141. What is the threshold velocity for particle movement in surface wind ?

(a) 8 km/hr (b) 3 km/hr
(c) 4 km/hr (d) None of these

142. In the land capability classification scheme, the total number of land capability is

(a) 4 (b) 6
(c) 5 (d) 8

143. The soil erodibility factor (K) indicates the inherent erodibility of a soil gives an indication of the soil loss from a unit plot _______ m long with 9% slope under continuous fallow culture.

(a) 12 (b) 22
(c) 72 (d) 40

144. Land capability class restricted to recreation wildlife, water supply or aesthetic purposes is

(a) VI (b) VIII
(c) II (d) VII

145. Hill soils are generally

(a) Acidic (b) Alkaline
(c) Neutral (d) None of these

146. The diagnostic surface horizons are called

(a) Pedons (b) Endopedons
(c) Genetic horizons (d) Epipedons

147. Available water holding capacity is maximum in

(a) Clay loam (b) Sand clay loam
(c) Silty clay loam (d) Loam

148. Size of colloidal clay particle is less than

(a) 0.0002 m (b) 0.02 mm
(c) 0.002 mm (d) None of these

149. Specific heat is higher in

(a) Silt (b) Humus
(c) Clay (d) Sand

150. Crop and cropping system types that can be used to conserve soil against soil erosion from region mainly depending on :

(a) Climate (b) Soil type
(c) Both (a) and (b) (d) None of these

Fill in the blanks

1. Muddy run-off from the field is an indication of ________ erosion.
2. Run-off and soil loss can be calculated by ________ .
3. The desertification was first used by ________ .
4. In North-eastern hill regions of India, shifting cultivation is also called as ________ .
5. Strong dusty winds are locally called as ________ .
6. In Chepil and Woodruff's equation of soil loss calculation by wind erosion, E =f (ICKLV), where K denotes ___________ .
7. Saltation occurs when particle size is _________ .
8. CSWCRTI was established in the year ___________ .
9. The crops which help to effectively check soil erosion and conserve soil moisture and nutrients are referred to as ________ .
10. The erosive power of rainfall, running water or wind is called _________ .
11. First recorded soil conservation research in India started in the year ________ .
12. Wind erosion is calculated by ___________ .
13. In land capability classification scheme, the class V, VI and VII are suitable for _____ .

14. First soil conservation scheme called dry farming scheme started at _______ in India.
15. Most common size of rain drop is ________ .
16. Erosion that can be easily removed by normal tillage operations is ________ .
17. Vegetative windbreaks are also called as ________ .
18. The volume of water that runs off during an irrigation that exceeds the soil's infiltrability is called ________ .
19. Vertical downward movement of water with in the soil is called _______ .
20. Poor water holding capacity of red soils is mainly due to _________ .

True/False

1. Water harvesting is defined as application of water at the time of harvesting crops like colocasia, tapioca and yam. **(T)**
2. At Kota, the best vegetative cover was provided by sunhemp. **(F)**
3. Fine textured soils are more susceptible to water erosion. **(T)**
4. In surface creep wind erosion, soil particles are moved primarily by the action of saltation. **(T)**
5. Removal by rain of a very thin layer of soil from the entire surface of large area is called Rill erosion. **(F)**
6. The ratio of dispersion ratio to colloidal moisture equivalent ratio is Erosion ratio. **(T)**
7. Scattering of detached soil particles by the impact of rain drops is called Splash erosion. **(T)**
8. Sheet erosion is observed on gently sloping land. **(T)**
9. The downward flow of water from the surface into the soil is Hydraulic conductivity. **(F)**
10. Clay content is higher in black soil. **(T)**

Answers Fill in the Blanks

1. Sheet
2. Ramser's formula
3. Abreville
4. Jhuming cultivation
5. Loo
6. Soil roughness
7. 0.1 – 0.5 mm
8. 1974
9. Cover Crops
10. Erosivity
11. 1923
12. Chepil and Woodruff equation
13. Pasture and Grazing
14. Manjri Near Pune
15. 7 mm
16. Rill erosion
17. Shelter belts
18. Run - off
19. Percolation
20. High infiltration

ANSWERS

1.(a)	2.(a)	3.(c)	4.(d)	5.(d)	6.(a)	7.(d)
8.(b)	9.(b)	10.(c)	11.(d)	12.(a)	13.(c)	14.(a)
15.(c)	16.(a)	17.(c)	18.(c)	19.(c)	20.(d)	21.(a)
22.(a)	23.(a)	24.(d)	25.(d)	26.(b)	27.(a)	28.(a)
29.(a)	30.(a)	31.(d)	32.(a)	33.(b)	34.(a)	35.(a)
36.(a)	37.(c)	38.(c)	39.(d)	40.(c)	41.(d)	42.(c)
43.(d)	44.(d)	45.(d)	46.(d)	47.(d)	48.(d)	49.(d)
50.(c)	51.(a)	52.(d)	53.(d)	54.(b)	55.(a)	56.(a)
57.(b)	58.(b)	59.(c)	60.(d)	61.(a)	62.(a)	63.(d)
64.(b)	65.(d)	66.(a)	67.(a)	68.(b)	69.(b)	70.(c)
71.(a)	72.(d)	73.(c)	74.(d)	75.(d)	76.(c)	77.(d)
78.(c)	79.(d)	80.(a)	81.(b)	82.(c)	83.(a)	84.(b)
85.(a)	86.(a)	87.(a)	88.(a)	89.(c)	90.(d)	91.(c)
92.(d)	93.(c)	94.(a)	95.(d)	96.(a)	97.(c)	98.(c)
99.(d)	100.(d)	101.(a)	102.(d)	103.(a)	104.(c)	105.(c)
106.(b)	107.(c)	108.(a)	109.(a)	110.(b)	111.(a)	112.(b)
113.(d)	114.(b)	115.(b)	116.(c)	117.(c)	118.(d)	119.(a)
120.(d)	121.(a)	122.(b)	123.(a)	124.(d)	125.(d)	126.(b)
127.(a)	128.(b)	129.(a)	130.(a)	131.(b)	132.(b)	133.(a)
134.(c)	135.(b)	136.(a)	137.(a)	138.(c)	139.(d)	140.(d)
141.(b)	142.(d)	143.(b)	144.(b)	145.(a)	146.(b)	147.(a)
148.(a)	149.(b)	150.(c)				

7

Manures and Fertilizers

1. **"Soil fertility and fertilizer" was written by**

   (a) Odum (b) M.M. Rai

   (c) Tisdale *et al.* (d) None of these

2. **__________ is the ratio of the increase in osmotic pressure produced by the fertilizer to that produced by the same weight of sodium nitrate ($NaNO_3$)**

   (a) Osmotic pressure (b) Nutrient index

   (c) Sodium index (d) Salt index

3. **In which country most of the N required for crop production through N_2 fixation by legumes**

   (a) India (b) Sri Lanka

   (c) Pakistan (d) New Zealand

4. **Which of the following pairs is not correctly matched**

Organic substances	:	*C/N ratio*
a. Soil micro-organisms		8:1
b. Soil organic matter		10:1
c. Barnyard manure (rotted)		20:1
d. Saw dust		1000:1

5. **The essentiality of iron to plants discovered by**

   (a) Gris (b) Justin von Liebig

   (c) Sprengel (d) Theodore de Saussure

6. **The essentiality of nitrogen to plants discovered by**

   (a) Gris (b) Sprengel

   (c) L. Sommer (d) Theodore de Saussure

7. **The essentiality of boron to plants discovered by**
   (a) Sprengel (b) L. Sommer and P. Lipman
   (c) K. Warrington (d) P. R. Stout

8. **The essentiality of molybdenum to plants discovered by**
   (a) Sprengel (b) P. Lipman
   (c) K. Warrington (d) I. Arnon and P. R. Stout

9. **The essentiality of zinc to plants discovered by**
   (a) Gris (b) Sprengel
   (c) L. Sommer and P. Lipman (d) None of these

10. **Primary plant nutrient is a**
   (a) Sulphur (b) Cobalt
   (c) Sodium (d) Potassium

11. **In the following phosphatic fertilizers which is called as concentrated superphosphate ?**
   (a) Single superphosphate (b) Double superphosphate
   (c) Triple superphosphate (d) All of these

12. **Monoammonium phosphate is prepared by the reaction of**
   (a) Single superphosphate with ammonia
   (b) Double superphosphate with ammonia
   (c) Triple superphosphate with ammonia
   (d) Both (a) and (b)

13. **The conversion factor for the conversion of MgO in to Mg is**
   (a) 0.608 (b) 0.39
   (c) 1.65 (d) None of these

14. **The conversion factor for the conversion of Ca in to CaO is**
   (a) 0.751 (b) 0.399
   (c) 0.608 (d) 1.658

15. **The conversion factor for the conversion of CaO in to Ca is**
   (a) 0.398 (b) 0.751
   (c) 1.129 (d) None of these

16. **In the process of denitrification, which gas largely escapes to the atmosphere?**
   (a) N_2O (b) SO_2
   (c) NO_2 (d) NH_3

17. **Which of the following fixes atmospheric nitrogen ?**
   (a) *Azotobacter* (b) *Nitrobacter*
   (c) *Clostridium* (d) *Pseudomonas*

18. Which one of the following micro-organism is responsible for nitrification ?

(a) *Nitrosomonas* (b) *Nitrobacter*
(c) *Pseudomonas* (d) *Azotobacter*

19. Conversion of ammonical nitrogen to nitrate nitrogen is termed as

(a) Ammonification (b) Nitrification
(c) Denitrification (d) Nitrate formation

20. Volatilization loss of nitrogen is comparatively more in

(a) Acid soils (b) Alkaline soils
(c) Neutral soils (d) None of these

21. The number of essential elements required for the growth of most higher plant is

(a) 7 (b) 5
(c) 6 (d) 17

22. Which of the following is basic fertilizer ?

(a) Urea (b) Calcium ammonium nitrate
(c) Sodium nitrate (d) Ammonium sulphate

23. Which fertilizer material is partly soluble in water ?

(a) Urea (b) Calcium ammonium nitrate
(c) Sodium nitrate (d) Magnesium sulphate

24. Which fertilizer is highly hygroscopic ?

(a) Urea (b) Calcium ammonium nitrate
(c) Sodium nitrate (d) Magnesium sulphate

25. In the following, which fertilizer contains sulphur ?

(a) Urea (b) Calcium ammonium nitrate
(c) Sodium nitrate (d) Ammonium sulphate

26. CDU, IBDU and Oxamide are

(a) Low water soluble compounds
(b) Sparingly water soluble compounds
(c) Soluble or relatively soluble compounds
(d) Urease inhibitors

27. The percentage of nitrogen in urea-formaldehyde is

(a) 40% (b) 30%
(c) 4% (d) 38%

28. The percentage of nitrogen in sulphur coated urea is

(a) 30% (b) 38%
(c) 21% (d) 46%

29. The percentage of nitrogen in thiourea is

(a) 30-32 % (b) 33-35%
(c) 21.8% (d) 36.8%

30. What is the percentage of nitrogen in raw bone meal ?

(a) 4 - 8 (b) 2 - 5
(c) 3 - 5 (d) 5 - 8

31. For nitrification, the optimum pH is

(a) 8.5 (b) 6.0
(c) 7.0 (d) 10.0

32. Which fertilizer has risk of fire or even explosion

(a) Aqua NH_3 (b) Ammonium nitrate
(c) Potassium nitrate (d) None of these

33. In agricultural soils, ______ of the total nitrogen is found as fixed ammonium ion

(a) 15% (b) 40%
(c) 10% (d) 35%

34. Zinc sulphate monohydrate contains Zn%

(a) 19 (b) 54
(c) 21 (d) 35

35. Chemical formula of pyrite reclaiment is

(a) ZnS (b) CuS
(c) MnS (d) FeS_2

36. The C:N ratio of organic matter is

(a) 2:8 (b) 10:1
(c) 19:1 (d) 16:1

37. The chemical formula of anhydrous ammonia is

(a) $NH_4.nH_2O$ (b) NH_4
(c) NH_3 (d) $NH_4.NO_3$

38. Enzyme involved in the hydrolysis of urea is the

(a) Urease (b) Nitrite reductase
(c) Nitrate reductase (d) Dehydrogenase

39. Triple superphosphate is prepared by the reaction of

(a) Rock phosphate with nitric acid
(b) Rock phosphate with acetic acid
(c) Rock phosphate with sulphuric acid
(d) Phosphoric acid with sulphuric acid

40. Single superphosphate is prepared by the reaction of

(a) Rock phosphate with sulphuric acid
(b) Rock phosphate with nitric acid
(c) Rock phosphate with phosphoric acid
(d) Phosphoric acid with sulphuric acid

41. What is the percentage of calcium in single super phosphate ?

(a) 35-40% (b) 33-35%
(c) 22-24% (d) 18-21%

42. Percentage of sulphur in triple phosphate is

(a) 1.0-1.5% (b) 5-8%
(c) 3-2.5% (d) None of these

43. Percentage of P_2O_5 in mono ammonium phosphate is

(a) 42-52% (b) 48-62%
(c) 43-44% (d) 12-18%

44. Percentage of P_2O_5 in urea ammonium phosphate is

(a) 13-14% (b) 48-60%
(c) 43-44% (d) None of these

45. Percentage of sulphate in urea sulphate is

(a) 46 - 53 % (b) 43 - 44%
(c) 6 - 11% (d) 18 - 23%

46. Which is the examples of macro nutrients ?

(a) N, P and K (b) Ca, Mg and S
(c) Fe, Mn and B (d) Both (a) and (b)

47. Fertilizers containing Cl ions are not suitable for

(a) Sugarcane (b) Potato
(c) Barley (d) Pea

48. Which is the stem nodulating green manure plant ?

(a) *Crotolaria juncea* (b) *Sesbania aculiata*
(c) *Sesbania rostrata* (d) None of these

49. A fodder and pasture legume

(a) Sugarcane (b) Stylosanthus
(c) Sesbania (d) None of these

50. Which one of the following oilcakes have highest nitrogen content ?

(a) Mustard cake (b) Neem cake
(c) Coconut cake (d) Groundnut cake

51. **What is the percentage of sulphur in potassium sulphate (K_2SO_4) ?**
(a) 21.0 (b) 18.5
(c) 22.0 (d) 12-14

52. **Percentage of sulphur in gypsum is**
(a) 21.0% (b) 23.5%
(c) 4% (d) 18.5%

53. **What is the percentage of sulphur in ammonium sulphate [NH_4 $(SO_4)_2$] ?**
(a) 23.8% (b) 18.5%
(c) 22.0% (d) 21.0%

54. **What is the percentage of calcium in calcium ammonium nitrate (CAN) ?**
(a) 18.5% (b) 8.0%
(c) 21% (d) 26.8%

55. **Percentage of calcium in gypsum is**
(a) 30.0% (b) 21%
(c) 18% (d) None of these

56. **Which fertilizer contains nutrients other than primary nutrients ?**
(a) Urea (b) Ammonium nitrate
(c) Ammonium sulphate (d) None of these

57. **What is the acidifying capacity of ammonium nitrate expressed in kg of $CaCO_3$ per kg of N supplied ?**
(a) 5.41 (b) 3.61
(c) 1.89 (d) 1.81

58. **Physiologically neutral fertilizer is**
(a) Calcium ammonium nitrate (b) Ammonium chloride
(c) Ammonium nitrate (d) None of these

59. **The chemical formula of calcium ammonium nitrate is**
(a) $(NH_4)_2 S_2 O_3$ (b) $Ca(H_2PO_4)_2$
(c) $CaCO_3.NH_4NO_3$ (d) $CaHPO_4$

60. **Percentage of Nitrogen in $Ca(NO_3)_2$**
(a) 14% (b) 15%
(c) 21% (d) 20%

61. **In magnesia (MgO) percentage of magnesium is**
(a) 45% (b) 55%
(c) 60% (d) 65%

62. **In basic slag percentage of magnesium is**
(a) 3-4% (b) 3-8%
(c) 20% (d) 26%

63. In basic slag percentage of P_2O_5 is

(a) 3-4% (b) 3-8%
(c) 20% (d) 26%

64. In basic slag the percentage of calcium is

(a) 28% (b) 20%
(c) 24% (d) 16%

65. In serpentine the percentage of magnesium is

(a) 26% (b) 40%
(c) 14% (d) None of these

66. The first fertilizer produced in India is

(a) Single superphosphate (b) Triple superphosphate
(c) Diammonium phosphate (d) None of these

67. The high fertilizer consuming state in India is

(a) U.P (b) Gujarat
(c) Haryana (d) Punjab

68. The desired ratio of NPK India is

(a) 2:1:1 (b) 3:2:1
(c) 4:2:1 (d) 3:1:2

69. What is the commonly used NPK ratio in India ?

(a) 8.9:2.7:1 (b) 8:3:1
(c) 9:3:1 (d) 4:2:2

70. Which is an example of fertilizer elements ?

(a) N, P and K (b) Ca, Mg and S
(c) Both (a) and (b) (d) None of these

71. In the following set of essential nutrient elements, which are absorbed as anions by plants ?

(a) Copper, iron, phosphorus, sulphur, zinc
(b) Potassium, manganese, molybdenum, phosphorus, zinc
(c) Boron, chloride, molybdenum, phosphorus, sulphur
(d) None of these

72. Which fertilizer supplies three essential plant nutrients ?

(a) DAP (b) MOP
(c) SSP (d) All of these

73. Which of the following iron chelate suitable for application to crops grown on acid soils ?

(a) Fe-EDTA (b) Fe-EDDHA
(c) Fe-HEDTA (d) None of these

74. Which of the following fertilizers contain water soluble phosphorus ?

(a) SSP
(b) MAP
(c) Both (a) and (b)
(d) None of these

75. In which of the following green manure crop stem nodulation occurs ?

(a) *Sesbania aculeate*
(b) *Sesbania rostrata*
(c) *Crotolaria juncea*
(d) None of these

76. Which one of the following is a green house gas ?

(a) Chlorine
(b) Ammonia
(c) Methane
(d) Hydrogen

77. What is the examples of secondary nutrients ?

(a) N, P and K
(b) Ca, Mg and S
(c) Fe, Mn and B
(d) All of these

78. Which is the examples of major nutrients ?

(a) N, P and K
(b) Ca, Mg and S
(c) Fe, Mn, CU, Cl, Mo and B
(d) All of these

79. How much minute of potash will be required to supply 50 kg K_2O/ha

(a) 150 kg
(b) 83.3 kg
(c) 120.3 kg
(d) 73.3 kg

80. Flame photometer is used in determination of

(a) Nitrogen
(b) Zinc
(c) Potassium
(d) Phosphorus

81. Sulphur content in ammonium sulphate is

(a) 24%
(b) 15%
(c) 30%
(d) 36%

82. Soil having 200 kg available K_2O/ha will be classed as

(a) Low
(b) Medium
(c) High
(d) Both (a) and (b)

83. Concentration of nitrogen in soil air is

(a) 75%
(b) 60%
(c) 92%
(d) 78%

84. Critical limit of available phosphorus in soil for wheat crop by Olsen's method is

(a) 5 kg/ha
(b) 27 kg/ha
(c) 10 kg/ha
(d) 15.7 kg/ha

85. Which of the following is correct

(a) $P_2O_5 \times 0.44 = P$
(b) $P_2O_5 \times 2.29 = P$
(c) $P_2O_5 \times 0.83 = P$
(d) $P_2O_5 \times 1.20 = P$

86. Widely used solid nitrogenous fertilizer is

(a) Ammonium sulphate (b) Ammonium nitrate
(c) Urea (d) None of these

87. Fertilizer would you prefer for groundnut grown on light soil is

(a) Ammonium sulphate (b) Urea phosphate
(c) Ammonium chloride (d) Ammonium sulphate nitrate

88. Fertilizer would you prefer for low land paddy is

(a) Urea (b) Ammonium nitrate
(c) Ammonium sulphate (d) None of these

89. The fertilizer that may cause maximum phytotoxicity is

(a) Urea phosphate (b) Urea
(c) Ammonium sulphate nitrate (d) Ammonium sulphate

90. Which is ammonium forming fertilizer ?

(a) Ammonium sulphate (b) Ammonium sulphate nitrate
(c) Calcium ammonium nitrate (d) Urea

91. Percentage of calcium in single super phosphate is

(a) 18-21% (b) 12-14%
(c) 25-29% (d) None of these

92. Percentage of sulphur in single super phosphate is

(a) 10% (b) 42%
(c) 12% (d) 24%

93. Percentage of P_2O_5 in single super phosphate is

(a) 40-52% (b) 36%
(c) 12% (d) 16%

94. Which nutrients increases the vigour and disease resistance to plants ?

(a) Nitrogen (b) Potassium
(c) Sulphur (d) None of these

95. Of the following nutrients, which one increases the winter hardiness, reduces lodging of crops ?

(a) Nitrogen (b) Potassium
(c) Sulphur (d) Ca and Mg

96. Which is essential in the formation and transfer of starches and sugars ?

(a) Nitrogen (b) Potassium
(c) Sulphur (d) None of these

97. Nutrient essential for stomatal regulations in plants is

(a) Calcium (b) Sulphur
(c) Potassium (d) Molybdenum

98. The purpose of nitrification inhibitors is

(a) To stop conversion nitrite to nitrate and hold nitrite by anion exchange to reduce its chances of leaching to ground water.
(b) To stop rapid mineralization of NH_4^+ to reduce leaching losses of N to ground water.
(c) To convert nitrate to NH_4^+ to reduce the potential losses of N as nitrate to ground water
(d) All of these

99. In S reduction, SO_4 to S^o is carried out by

(a) Desulphovibrio (b) Desulphomaculatum
(c) Both (a) and (b) (d) None of these

100. What is the percentage of nitrogen in $CaCN_2$?

(a) 24.0% (b) 21.0%
(c) 43% (d) 35.0%

101. What is the conversion factor for the conversion of Mg in to MgO?

(a) 0.851 (b) 0.499
(c) 1.658 (d) 0.608

102. Which nutrients element is the basic constituent of life ?

(a) Nitrogen (b) Phosphorus
(c) Potassium (d) None of these

103. Which nutrient is essential for the increases in protein content in food fodder crops ?

(a) Nitrogen (b) Phosphorus
(c) Molybdenum (d) Potassium

104. Nutrients stimulates flowering and aids in seed form is

(a) Phosphorus (b) Potassium
(c) Nitrogen (d) Molybdenum

105. Nutrients, when applied to legumes, enhances the activity of rhizobia and increases the formation of root nodules is

(a) Phosphorus (b) Sulphate
(c) Potassium (d) Nitrogen

106. Which of the following elements, when excess may cause deficiencies of trace elements, particularly iron and zinc ?

(a) Phosphorus (b) Nitrogen
(c) Potassium (d) Sulphate

107. Treating rock phosphate with sulphuric acid for the production of phosphoric acid is called as

(a) Green process (b) Wet process
(c) Both (a) and (b) (d) None of these

108. Single super phosphate is

(a) A neutral fertilizer (b) An acid producing fertilizer
(c) A alkali producing fertilizer (d) All of these

109. What is the percentage of citrate soluble P_2O_5 in rock phosphate ?

(a) 25-30% (b) 5-17%
(c) 28-36% (d) 13-18%

110. The product of the reaction $NH_3 + H_4P_2O_7$ is

(a) Monoammonium phosphate (b) Dicalcium phosphate
(c) Triammonium phosphate (d) None of these

111. Who first discovered that plants need mineral nutrients for their growth and developments ?

(a) Nicholas Theodore de Saussure (b) Jean Baptiste Bousingault
(c) Sprengel (d) Both (a) and (b)

112. Which one of the following helps in absorption of nitrogen ?

(a) Mn (b) B
(c) Mo (d) None of these

113. Which of the following element is a constituent of cell membrane and in essential for cell division ?

(a) Potassium (b) Boron
(c) Molybdenum (d) Zinc

114. Which of the following element is essential for translocation of sugars in plant ?

(a) Boron (b) Molybdenum
(c) Zinc (d) Potassium

115. Which of the following element is essential for flowering and fruiting process and pollen germination ?

(a) Boron (b) Potassium
(c) Zinc (d) Magnesium

116. Which of the following element is not a constituent of chlorophyll but, it helps in its formation ?

(a) Fe (b) Mn
(c) Cu (d) None of these

117. In triple super phosphate calcium percentage is

(a) 12-14% (b) 24-28%
(c) 51-66% (d) 3-8%

118. Percentage of P_2O_5 in triple super phosphate is

(a) 12-14% (b) 18-21%
(c) 48-52% (d) None of these

119. Percentage of Calcium in $Ca(NO_3)_2$ is

(a) 19% (b) 49%
(c) 36% (d) 30%

120. The percentage of magnesium in $Mg(NO_3)_2$ is

(a) 24% (b) 16%
(c) 21% (d) 11%

121. Single super phosphate is best suitable for

(a) Neutral and alkaline soils (b) Acid soil only
(c) Alkaline soil only (d) Neutral soil only

122. A potential inhibitor of nitrification is

(a) CS_2 (b) Nitrogen
(c) ST (d) U-form

123. What is the percentage of nitrogen in ammonium thiosulphate[$(NH_4)_2S_2O_3$]?

(a) 24% (b) 18%
(c) 12% (d) 28%

124. What is the percentage of sulphur in ammonium thiosulphate[$(NH_4)_2S_2O_3$]?

(a) 26% (b) 18%
(c) 12% (d) 21%

125. Which of the following is not a constituent of chlorophyll but, it helps in chlorophyll formation and encourages vegetative plant growths ?

(a) Phosphorus (b) Potassium
(c) Sulphur (d) Nitrogen

126. Which of the following is an essential constituent of certain volatile compounds and helps in the reduction-oxidation system in respiration ?

(a) Phosphorus (b) Potassium
(c) Sulphur (d) Nitrogen

127. The element essential for the calcium metabolism, both with its uptake by roots and its efficient use in plants is

(a) Magnesium (b) Phosphorus
(c) Nitrogen (d) Boron

128. Nutrient element influences the formation of some growth hormones in the plants is

(a) Cu (b) B

(c) Zn (d) Mn

129. Nutrient element brings about the reduction of nitrates to ammonia prior to amino acid and protein synthesis in the cells of the plant is

(a) Iron (b) Molybdenum

(c) Zinc (d) Potassium

130. Which element is essential for nitrogen fixation, both (symbiotic and non-symbiotic) ?

(a) Boron (b) Molybdenum

(c) Zinc (d) Potassium

131. Which of the following elements acts as "electron carrier" in enzymes which brings about oxidation-reduction reactions in plants ?

(a) Zinc (b) Iron

(c) Boron (d) Copper

132. Which of the following is the acid soluble phosphatic fertilizer ?

(a) Super phosphate (b) Monocalcium phosphate

(c) Rock phosphate (d) None of these

133. In DAP, the percent content of N and P is

(a) 20 - 20 (b) 18 - 46

(c) 46 - 18 (d) 22 - 48

134. The water soluble phosphate is

(a) $Ca\ (H_2PO_4)_2$ (b) $CaHPO_4$

(c) $Ca_3\ (PO_4)_2$ (d) $CaCO_3 . NH_4 NO_3$

135. The conversion factor to convert % P to % P_2O_5 is

(a) 1.29 (b) 2.90

(c) 2.29 (d) 1.80

136. Which of the following the fertilizer contains maximum sulphur ?

(a) Nitric phosphate (b) Concentrated super phosphate

(c) Single super phosphate (d) Enriched super phosphate

137. The super phosphate is largely produced, now a days from

(a) Phosphate guanos (b) Rock phosphate

(c) Bone meal (d) Basic slag

138. In KCl the per cent of K_2O is

(a) 88 (b) 64

(c) 124 (d) 70

139. Which of the following elements helps in utilization of iron in chlorophyll synthesis ?

(a) Boron (b) Zinc
(c) Iron (d) Copper

140. Which of the following elements are classified on the functional basis as accessory structural elements ?

(a) C and H (b) C, H and O
(c) N and P (d) None of these

141. Which of the following elements is classified on the functional basis as accessory structural elements ?

(a) C (b) H
(c) P (d) S

142. Which of the following elements are classified on the functional basis as regulators and carriers ?

(a) C, H, O (b) N, P, S
(c) K, Ca, Mg (d) None of these

143. Which of the following elements are classified on the functional basis as main structure elements ?

(a) C, H, O (b) N, P, S
(c) K, Ca, Mg (d) None of these

144. An essential nutrient element is

(a) Barium (b) Iodine
(c) Boron (d) None of these

145. An organic chemical fertilizer is

(a) Calcium ammonium phosphate (b) Urea
(c) Ammonium phosphate (d) None of these

146. A complex fertilizer is

(a) Ammonium sulphate nitrate (b) Calcium ammonium nitrate
(c) Single superphosphate (d) None of these

147. Of the following, which is an ultra micro-nutrient ?

(a) Zinc (b) Molybdenum
(c) Iron (d) None of these

148. In urea, nitrogen per cent is

(a) 20 - 21 (b) 33 - 34
(c) 26 - 30 (d) 44 - 46

149. The permissible maximum biuret content of urea produced in India is ______ per cent

(a) 2.0 (b) 3.0
(c) 1.5 (d) 1.0

150. Single superphosphate is produced by the reaction of rock phosphate with

(a) HNO_3 (b) H_2O_2
(c) H_2SO_4 (d) None of these

151. Concentrated super phosphate is produced by the reaction of rock phosphate with

(a) HCl (b) H_3PO_4
(c) H_2SO_4 (d) H_2O_2

152. The chief component of single superphosphate that supplies phosphorus to plant is

(a) Rock phosphate (b) Free phosphoric acid
(c) Monocalcium phosphate (d) None of these

153. The potassium meta phosphate is

(a) K_2O (b) KH_2PO_4
(c) KPO_3 (d) K_2HPO_4

154. Which form of nitrogen absorbed by plant is toxic ?

(a) NH_4+ (b) NO_3^-
(c) NO_2^- (d) None of these

155. Ammonia volatilization is highest in

(a) $CaCO_3 . NH_4 NO_3$ (b) $NH_4 NO_3$
(c) $CaCO_3 . NH_4 NO_3$ (d) $(NH_4)_2SO_4$

156. The process in which nitrate is reduced is known as

(a) Chemo-denitrification (b) Indirect denitrification
(c) Biological nitrification (d) Denitrification

157. The predominant phosphate in single superphosphate is

(a) Tricalcium phosphate (b) Dicalcium phosphate
(c) Monocalcium phosphate (d) None of these

158. Root development is stimulated due to supply of

(a) Zn (b) P
(c) N (d) B

159. Fixation of phosphate is more in

(a) Acid soils (b) Neutral soils
(c) Submerged soils (d) None of these

160. Trench method of preparing FYM has been given

(a) M. S. Swaminathan (b) Wager and Krosender
(c) S. V. Desai (d) N. Acharya

161. Which of the following elements are classified on the functional basis as catalysts and activators?

(a) Fe, Mn, Zn , Cu (b) B, Mo, Cl
(c) Fe, Mn, Cu, Mo (d) All of these

162. Nutrient element is essential for the synthesis and translocation of carbohydrates in plant is

(a) K (b) Ca
(c) Mg (d) Fe

163. Which one of the following nutrients loss is most by leaching ?

(a) N (b) P
(c) K (d) Zn

164. Which of the following is a micro-nutrient fertilizer ?

(a) Zinc sulphate (b) Magnesium sulphate
(c) Potassium sulphate (d) None of these

165. Biuret in urea applied as a foliar spray in citrus fruits should not be more than ?

(a) 0.25% (b) 0.50%
(c) 1.75% (d) 1.50%

166. Which fertilizer act as a nitrification inhibitor and as a slow release N source

(a) CDU (b) IBDU
(c) N-serve (d) DCD

167. The inorganic and organic labile P fraction of soil are collectively called

(a) Intensity factor (b) Quantity factor
(c) Capacity factor (d) None of these

168. Select the pair which is not correctly matched

C/P ratio	:	***Mineralization/immobilization***
a. < 200	:	Net mineralization of organic P
b. 200-300	:	No gain/loss of inorganic P
c. >300	:	Net immobilization of inorganic P
d. > 400	:	Net immobilization of organic P

169. Fluor apatite is rich source of

(a) Phosphorus (b) Molybdenum
(c) Nitrogen (d) Potassium

170. Amongst oil cakes, the highest nitrogen content is in

(a) Castor cake (b) Neem cake
(c) Coconut cake (d) Groundnut cake

171. The readily available form of K in soil ranges from

(a) 0.1-2.0% (b) 90-98%
(c) 2-5% (d) 40-60%

172. P contents in DAP is

(a) 50% (b) 60%
(c) 46% (d) 20%

173. Chemical formulae of single super phosphate is

(a) $Ca(H_2PO_4)$ (b) $CaH\ PO_4$
(c) $Ca_3\ (PO_4)$ (d) $Ca(H_2PO_4).\ CaSO_4.\ 2H_2O$

174. In India, which fertilizer manufactured first

(a) DAP (b) ASN
(c) Urea (d) SSP

175. End product of denitrification process is

(a) N_2 (b) NH_4
(c) NO_3 (d) None of these

176. Which nitrogenous fertilizer industrially made first ?

(a) $NaNO_3$ (b) $CaCN_2$
(c) Ammonium sulphate (d) Urea

177. Biosuper is a fertilizer containing

(a) Rock phosphate + elemental sulphur + S. oxidizing bacteria
(b) Phosphate acid + elemental sulphur + S. oxidizing bacteria
(c) Ammonium sulphate + elemental sulphur + S. oxidizing bacteria
(d) None of these

178. Fertilizer control order issued in

(a) 1986 (b) 1957
(c) 1976 (d) 1986

179. Application of potash increases

(a) Resistance to water logging (b) Frost resistance in plants
(c) Disease resistance in plants (d) None of these

180. Biofertilizers can be used to achieve

(a) To fix atmospheric nitrogen
(b) To solubilize soil nutrient like phosphate
(c) To stimulate plant growth
(d) All of these

Fill in the Blanks

1. Concentrated organic manures are also classified as __________ .
2. C:N ratio of normal soil is _________ .
3. The fertilizer applied at the time of sowing is _______ application, while that is applied to standing crop is referred to _______ .
4. The fixation of "P" is minimum at pH ________ .
5. Fertigation technique increased the fertilizer use efficiency by ________ .
6. Dhaincha is ______ green manure crop.
7. Urea is _______ fertilizer.
8. Urea in which oxygen is substituted by urea is _______ .
9. Paddy crop takes the nitrogen in the form of ________ .
10. Carbon-nitrogen ratio of normal soil is ________ .
11. Nitrogen content in green leaves of glyricidia is ________ .
12. Organic carbon content in organic matter is ________ .
13. Nitrogen content in neem cake is _______ .
14. Nitrogen loss from FYM is prevented by ________ .
15. Biuret percent in urea fertilizer is _________ .
16. Nitrogen percent in urea is _________ .
17. Ammonium sulphate is preferred for nitrogen application in groundnut because it also provides _________ .
18. Potassic fertilizer is used for improving quality of fruits is __________ .
19. Single super phosphate is produced by the reaction of rock phosphate with ______ .
20. The chief component of single superphosphate that supplies phosphorus to plant is _______.

True/False

1. Diammonium phosphate is complex fertilizer.**(T)**
2. Azolla is used as green manure in rice. **(T)**
3. Nitrate fertilizer is the carbon containing fertilizer. **(F)**
4. Sulphate is the available form of sulphur in soil. **(T)**
5. Farm yard manure recommended for salinity reclamation. **(T)**
6. Biological nitrogen fixation is anaerobic. **(F)**
7. Denitrification is more in waterlogged soils. **(T)**
8. Murate of potash contains 36% K_2O. **(F)**

9. Urea is an organic chemical fertilizer. **(T)**
10. Ca $(H_2PO_4)_2$ is the water soluble phosphate. **(T)**
11. The number of essential elements required for the growth of most higher plant is 6. **(F)**
12. In DAP the per cent content of N and P is 18 - 46. **(T)**
13. In India, Punjab is the higher fertilizer consuming state. **(T)**
14. Nitrosomonas fixes atmospheric nitrogen. **(F)**
15. Nitrification is the conversion of ammoniacal nitrogen to nitrite nitrogen. **(T)**

Answers Fill in the Blanks

1. Nitrogenous Fertilizer
2. 10:1
3. Leguminaceae, Gramineae
4. 6.0–7.0
5. 50 – 60%
6. Leguminous
7. Organic
8. Thiourea
9. Amide
10. 10:1 to 12:1
11. 3.0%
12. 58%
13. 5.2 %
14. Gypsum
15. 1.5 %
16. 46 %
17. Sulphur
18. Single Super Phosphate
19. Sulphuric acid (H_2SO_4)
20. Monocalcium phosphate

ANSWERS

1.(c)	2.(d)	3.(d)	4.(d)	5.(a)	6.(d)	7.(c)
8.(d)	9.(c)	10.(d)	11.(c)	12.(d)	13.(a)	14.(b)
15.(b)	16.(a)	17.(a)	18.(b)	19.(b)	20.(b)	21.(d)
22.(c)	23.(b)	24.(a)	25.(d)	26.(a)	27.(d)	28.(b)
29.(d)	30.(c)	31.(a)	32.(b)	33.(c)	34.(d)	35.(d)
36.(b)	37.(c)	38.(a)	39.(d)	40.(a)	41.(d)	42.(a)
43.(b)	44.(a)	45.(c)	46.(d)	47.(b)	48.(c)	49.(b)
50.(d)	51.(b)	52.(a)	53.(a)	54.(b)	55.(a)	56.(c)
57.(c)	58.(a)	59.(c)	60.(b)	61.(b)	62.(a)	63.(b)
64.(b)	65.(a)	66.(a)	67.(d)	68.(c)	69.(a)	70.(a)
71.(c)	72.(c)	73.(a)	74.(c)	75.(b)	76.(c)	77.(b)
78.(a)	79.(d)	80.(c)	81.(a)	82.(b)	83.(d)	84.(c)
85.(a)	86.(c)	87.(a)	88.(a)	89.(b)	90.(d)	91.(a)
92.(c)	93.(d)	94.(b)	95.(b)	96.(b)	97.(c)	98.(b)
99.(c)	100.(b)	101.(c)	102.(a)	103.(a)	104.(a)	105.(a)
106.(a)	107.(c)	108.(a)	109.(b)	110.(c)	111.(d)	112.(d)
113.(b)	114.(a)	115.(a)	116.(a)	117.(a)	118.(c)	119.(a)
120.(b)	121.(a)	122.(a)	123.(c)	124.(a)	125.(c)	126.(c)
127.(d)	128.(c)	129.(b)	130.(b)	131.(d)	132.(c)	133.(b)
134.(a)	135.(c)	136.(c)	137.(b)	138.(b)	139.(d)	140.(c)
141.(d)	142.(c)	143.(a)	144.(c)	145.(b)	146.(b)	147.(b)
148.(d)	149.(c)	150.(c)	151.(b)	152.(c)	153.(c)	154.(c)
155.(d)	156.(d)	156.(c)	157.(c)	158.(b)	159.(a)	160.(d)
161.(d)	162.(a)	163.(a)	164.(a)	165.(a)	166.(d)	167.(b)
168.(d)	169.(a)	170.(d)	171.(a)	172.(d)	173.(a)	174.(d)
175.(d)	176.(b)	177.(a)	178.(b)	179.(c)	180. (d)	

8

Cropping Systems

1. **What is the basic principal of taking crop rotation ?**
   (a) To get higher crop production
   (b) To get higher returns per unit area of the soil
   (c) To maintain the fertility status of the soils
   (d) None of these
2. **Inter-cropping is a system of cropping to make the maximum use of**
   (a) Solar energy (b) Water resource
   (c) Soil resource (d) All of these
3. **What is relay cropping ?**
   (a) A system of cropping where one crop hands over the land to the next crop in a quick succession
   (b) A system of cropping where two or three crops are grown annually
   (c) A system of cropping where two or three crops are grown together by mixing the seeds.
   (d) All of these
4. **Practicing agriculture without the use of chemicals and fertilizers so as to protect environmental balance is known as**
   (a) Intensive farming (b) Natural farming
   (c) Inorganic farming (d) None of these
5. **The most effective cropping system for returning mineral elements to the soil is known as**
   (a) Relay cropping (b) Crop rotation
   (c) Double cropping (d) Strip cropping
6. **Which pair is wrong ?**
   (a) C_3-Maize (b) Calvin cycle - PGA
   (c) Hatch and Slack cycle - OAA (d) C_4- Kranz anatomy

7. **How many number of crops taken in relay cropping ?**
   (a) 2 (b) 3
   (c) 4 (d) 6

8. **Example of relay cropping is**
   (a) Maize - Potato - Cucurbit
   (b) Paddy - Wheat
   (c) Maize - Wheat
   (d) Soyabean - Sugarcane

9. **Rice - rice - rice rotation indicates**
   (a) Polyculture (b) Aquaculture
   (c) Monoculture (d) None of these

10. **The ratio of leaf area to covered ground area is known as**
    (a) Leaf area index (b) Leaf area ratio
    (c) Leaf area duration (d) Crop growth rate

11. **Practise of raising forest on the society land to minimise atmospheric pollution is known as**
    (a) Farm forestry (b) Silvipasture
    (c) Agroforestry (d) Social forestry

12. **Who was developed concept of agro-climatic zoning ?**
    (a) World bank (b) USA
    (c) FAO (d) ICAR

13. **What is lay farming ?**
    (a) Growing of field crops in rotation with grasses and fodder legumes
    (b) Growing trees in rotation with grasses and fodder legumes
    (c) Growing of pastures in rotation with grasses and fodder legumes
    (d) None of these

14. **What is alley cropping ?**
    (a) Growing of pastures in between two widely spaced rows of fast growing trees
    (b) Growing of field crops in between two widely spaced rows of fast growing trees/shrubs
    (c) Growing of only short duration crops in between two widely spaced rows of fast growing tree
    (d) None of these

15. **The intensity of paddy - pea - sugarcane - fallow - lobia- wheat would be**
    (a) 250% (b) 150%
    (c) 166.6% (d) None of these

16. **Taking two crops in a year in sequence from the same piece of land is known as**
    (a) Mono cropping (b) Double cropping
    (c) Quadruple cropping (d) Multiple cropping
17. **Growing three crops in a year in sequence is known as**
    (a) Mono cropping (b) Double cropping
    (c) Triple cropping (d) Multiple cropping
18. **Growing four crops in a year in sequence is known as**
    (a) Mono cropping (b) Double cropping
    (c) Quadruple cropping (d) None of these
19. **Growing two or more crops consecutively on the same field in the same year is known as**
    (a) Multiple cropping (b) Diversified cropping
    (c) Relay cropping (d) Double cropping
20. **Measures the sum of area planted to different crops and harvested in single year divided by the total cultivated area times 100 is called**
    (a) Multiple cropping index (b) Diversified index
    (c) Competition index (d) None of these
21. **System of growing together, crops of different heights at the same time on the same piece of land and thus using land, water & space most efficiently and economically is known as :**
    (a) One tier system (b) Two tier system
    (c) Three tier system (d) Multi tier system
22. **Cultivation of crop regrowth after harvest is known as**
    (a) Sequential cropping (b) Double cropping
    (c) Multiple cropping (d) Ratoon cropping
23. **Growing two or more crops simultaneously with no distinct row agreement is known as**
    (a) Mixed inter-cropping (b) Relay inter-cropping
    (c) Strip inter-cropping (d) None of these
24. **Growing two or more crops simultaneously with one or more crops in distinct row agreement is known as**
    (a) Mixed inter-cropping (b) Row inter-cropping
    (c) Relay inter-cropping (d) None of these
25. **Growing two or more crops simultaneously during part of life cycle of each is known as**
    (a) Relay inter-cropping (b) Mixed inter-cropping
    (c) Strip inter-cropping (d) None of these

26. A farming system in which arable crops are grown with trees or shrubs established mainly to hasten soil fertility restoration and enhance soil productivity is known as

(a) Alley cropping (b) Agro-forestry
(c) Lay farming (d) Relay cropping

27. A system in which land is managed for the concurrent production of agricultural and forest crops and for the rearing of domesticated animal

(a) Agro-silvi-pastoral system (b) Agri-horti-silvi pastoral system
(c) Horti-silvi-pastoral system (d) Both (b) and (c)

28. Agro-silvi-patoral system is a combination of agrisilviculture and

(a) Silvi-pastoral system (b) Agri-horti system
(c) Horti pastoral system (d) None of these

29. In which state, Jowar and Paddy are grown as mixed crops ?

(a) Uttar Pradesh (Rohilkhand) (b) Madhya Pradesh (Raipur)
(c) Haryana (Rohtak) (d) Tamil Nadu (Coastal region)

30. Individual crop species which are part of the multiple cropping system are known as

(a) Component crops (b) Multiple crops
(c) Companion crops (d) Mono crop

31. Which is the example of simultaneous poly culture ?

(a) Mixed cropping (b) Inter-cropping
(c) Relay planting (d) All of these

32. The complementary interaction between component crops both in space and time is termed as

(a) Allelopathy (b) Annidation
(c) Toxic effect (d) All of these

33. In multiseed cropping the principal of ________ is used

(a) Allelopathy (b) Annidation in space
(c) Annidation in time (d) Both (b) & (c)

34. Planting of shade trees in coffee plantation is an example of

(a) Allelopathy (b) Annidation in space
(c) Annidation in time (d) None of these

35. RYT is mostly used to measure yield advantage in which type of land ?

(a) Pasture land (b) Fallow land
(c) Crop land (d) None of these

36. In relay cropping

(a) Second crop is sown after the harvest of previous crop
(b) Second crop is sown before the harvest of previous crop
(c) Second crop is sown next year
(d) None of these

37. Wheat and barley are not grown as mixed crops in the state of

(a) Jammu and Kashmir (b) Rajasthan and M.P.
(c) Haryana (d) U.P. and Bihar

38. Cotton is grown as a mixed crop with rice in

(a) Gujarat (b) Maharashtra
(c) Assam (d) All of these

39. What are the important crop mixtures grown in Gujarat ?

(a) Jowar and tur or urad, bajra and moong or tur
(b) Wheat and gram, paddy and tur or cotton, maize and tur
(c) Both (a) and (b)
(d) None of the above

40. What are important crop mixtures grown in Maharashtra ?

(a) Jowar with tur or moth or safflower or moong or urad or cotton or groundnut or coconut or and tur or groundnut and tur or safflower and wheat or gram, rabi jowar with gram or moong
(b) Bajra with moth or castor or moth and hulga or tur or urad or niger or moth and moong
(c) Both (a) and (b)
(d) None of these

41. What is cropping intensity ?

(a) The extent of the use of land for purposes other than cropping
(b) The extent of the use of land for cropping purposes during a given year
(c) Gross cropped area
(d) None of these

42. Which crops are grown during summer ?

(a) *Rabi* (b) *Kharif*
(c) *Zaid* (d) None of these

43. In a sub-tropical area if the average rainfall is 400-500mm, which one of the following crop is most suitable to cultivate ?

(a) Rice (b) Pea
(c) Sorghum (d) Potato

44. Mixed farming referred only to which of the following ?

(a) Growing of crop together by mixing the seeds
(b) Growing of cash crops and food grains together
(c) Crop production combined with dairy farming at farm
(d) Growing of fruits and vegetable

45. What is cropping intensity ?

(a) Percentage ratio of gross cropped area to net cropped area
(b) Percentage ratio of net cropped area to gross cropped area
(c) Percentage ratio of the number of crops in rotation to period of one rotation
(d) None of these

46. A companion crop which nourishes the main crop by way of nitrogen fixation and/or adding the organic matter in to the soil is known as :

(a) Nurse crop (b) Guard crop
(c) Catch crop (d) None of these

47. Growing main crop in the centre surrounded by hardy and thorny crop such as safflower around pea, mesta around sugarcane and sorghum around maize is known as

(a) Guard crop (b) Truck crop
(c) Catch crop (d) None of these

48. The practice of placing primary emphasis on the use of stored soil moisture while determining whether or not to establish a crop is called :

(a) Multiple cropping (b) Diversified cropping
(c) Relay cropping (d) Opportunity cropping

49. The range of environmental factors over which an organism or a physiological process can function is known as

(a) Ecological amplitudes (b) Ecological equivalence
(c) Ecological pyramid (d) None of these

50. The situation in which two or more species can replace one another in the same ecological niche because of the similarity in their ecological amplitude is known as

(a) Ecological amplitudes (b) Ecological equivalence
(c) Ecological pyramid (d) Eco-niche

51. Crops which are grown during the two years is called :

(a) Annual crops (b) *Rabi* crops
(c) *Zaid* crops (d) Biennial crops

52. Crops which are grown during the more than two years is called :

(a) *Kharif* crops (b) Biennial crops
(c) *Zaid* crops (d) Perennial crops

53. The emergence of new tillers from the root/crown of a plant specially as in sugarcane, sorghum etc. is known as

(a) Ratoon (b) Precision farming
(c) Rangeland management (d) Recession farming

54. The cultivation of an additional crop from the regrowth of stubbles of previous main crop after its harvest, thereby avoiding reseeding/replanting is known as

(a) Contour farming (b) Ratoon cropping
(c) Rangeland management (d) Precession farming

55. Management of a range to produce maximum forage for sustained use without jeopardizing the other resources or uses of land is known as

(a) Contour farming (b) Recession farming
(c) Rangeland management (d) None of these

56. What is the best example of synergistic cropping system ?

(a) Pigeonpea + Mustard
(b) Potato + Wheat
(c) Autumn planted sugarcane + mustard
(d) Autumn planted sugarcane + potato

57. The plant geometry of wheat crop is :

(a) Square (b) Spherical
(c) Extreme rectangular (d) None of these

58. How many number of steps taken in preparation of cropping scheme ?

(a) 4 (b) 8
(c) 6 (d) 12

59. Yearly sequence and spatial arrangement of crops on a given land area is known as :

(a) Cropping system (b) Cropping pattern
(c) Sequential cropping (d) Mixed cropping

60. Crop should be harvested at :

(a) Physiological maturity stage (b) Real maturity stage
(c) Over maturity stage (d) None of these

61. Sowing of toria in standing crop of maize is :

(a) Mixed cropping (b) Mixed inter-cropping
(c) Alley cropping (d) Relay cropping

62. The highest yield of potato is obtained when cultivated in the rotation :

(a) Maize - potato - pea (b) Maize - potato - sunflower
(c) Dhaincha - potato - sunflower (d) None of these

63. Which value can be assessed by using equation

$$= \frac{\text{Total no. of crops} \times \text{Their respective area} \times 100}{\text{Total area available for cropping pattern}}$$

(a) Diversity index (b) Simultaneous cropping index
(c) Multiple cropping index (d) None of these

64. What is the intensity of groundnut + arhar + sugarcane crop rotation ?

(a) 150% (b) 200%
(c) 300% (d) 400%

65. What is alley cropping ?

(a) Growing of pastures in between two widely spaced rows fast growing trees
(b) Growing of field crops in between two widely spaced rows fast growing trees
(c) Growing of only short duration crops in between two widely spaced rows of fast growing trees
(d) Growing of only fodder crops in between two widely spaced rows of fast growing trees

66. In India crop seasons are named as

(a) *Kharif* season (b) *Zaid, Kharif* season
(c) *Rabi, Zaid* season (d) All of these

67. From being of May to middle of October period is

(a) *Kharif* season (b) *Zaid, Kharif* season
(c) *Rabi, Zaid* season (d) None of these

68. From August to January is the season of

(a) *Kharif* (b) *Zaid, Kharif*
(c) *Rabi, Zaid* (d) *Rabi*

69. Maize-toria-wheat-moong rotation is termed as a system of:

(a) Mixed cropping (b) Relay cropping
(c) Multiple cropping (d) None of these

70. Growing plants of different heights in the same field at the same time is known as :

(a) Multiple cropping (b) Multistoreyed cropping
(c) Mixed cropping (d) None of these

71. Planting of succeeding crop before harvest of standing crop is termed as :

(a) Mixed cropping (b) Relay cropping
(c) Inter cropping (d) None of these

72. **Soil fertility is reduced due to:**

(a) Continuous cropping (b) Imbalanced use of fertilizers
(c) Over irrigation (d) All of these

73. **In sugarcane + wheat inter-cropping, sugarcane is planted in the month of :**

(a) October (b) July
(c) March (d) May

74. **The intensity of jowar + arhar - fallow - wheat would be :**

(a) 250% (b) 150%
(c) 200% (d) None of these

75. **The intensity of fallow - sugarcane - rotton would be**

(a) 66.6% (b) 100%
(c) 150% (d) None of these

76. **The intensity of jowar + gaur - berseem + oat would be**

(a) 100% (b) 150%
(c) 200% (d) None of these

77. **The intensity of sunhemp (GM) - sugarcane would be**

(a) 100% (b) 75%
(c) 50% (d) None of these

78. **Who proposed multiple cropping index ?**

(a) Dalrimpal (1971) (b) Rao and Willey (1980)
(c) Monge and Associates (1978) (d) None of these

79. **Intensification of cropping both in time and space takes place in which one of the following?**

(a) Multiple cropping (b) Double cropping
(c) Sequential cropping (d) Mono cropping

80. **Intensification of cropping in time dimensions takes place in which one of the following ?**

(a) Multiple cropping (b) Double cropping
(c) Sequential cropping (d) Mono cropping

81. **Crop mixtures not important for farmers of Himachal Pradesh are**

(a) Wheat and barley, barley and gram
(b) Wheat and gram, maize and sesamum
(c) Wheat and mustard, maize and pulses
(d) None of these

82. **Intensification of cropping both in time and space takes place in which one of the following ?**
(a) Multiple cropping (b) Inter-cropping
(c) Mono-cropping (d) Both (a) and (b)

83. **Which one of the following is the most important crop mixture grown by farmers in Tripura ?**
(a) Cotton, sessamum and autumn paddy
(b) Paddy and chillies
(c) Wheat and peas
(d) Wheat and mustard

84. **In the following crop mixtures which are not important for farmers of Manipur ?**
(a) Rice and chillies
(b) Cotton, sessamum and autumn rice
(c) Wheat and peas
(d) Wheat and mustard

85. **The yearly sequence and spatial arrangement of crops and fallow in an area is termed as**
(a) Cropping pattern (b) Cropping system
(c) Crop rotation (d) Farming pattern

86. **The recurrent succession of crops on the same piece of land either in a year or over a period of time is known as**
(a) Farming system (b) Cropping system
(c) Crop rotation (d) Inter cropping

87. **What is the main objective of mixed farming ?**
(a) Subsistence (b) Higher profitability
(c) Ecological balance (d) All of these

88. **What is the main objective of farming system ?**
(a) Subsistence (b) Higher profitability
(c) Ecological balance (d) Both (b) & (c)

89. **The proportion of area under different crops at a point of time in a unit area is known as**
(a) Cropping pattern (b) Cropping system
(c) Crop rotation (d) Inter cropping

90. **In low rainfall area i.e. < 750 mm /annual which type of cropping is practised ?**
(a) Mono cropping (b) Double cropping
(c) Multiple cropping (d) Inter-cropping

91. Growing of crops one after the other without the seasonal following is known as

(a) Continuous cropping (b) Contingency cropping
(c) Contour cropping (d) Relay cropping

92. Growing of crops in aberrant situations like drought and flood is known as

(a) Continuous cropping (b) Contingency cropping
(c) Contour cropping (d) None of these

93. The cultivation of crops along the contours of a slope is known as

(a) Conservation cropping (b) Contingency cropping
(c) Contour cropping (d) None of these

94. As the ability to regenerate by shoot or root suckers, or a forest to established is known as

(a) Coppice farming (b) Cover crop
(c) Crop competition (d) None of these

95. The crops which are grown around the field boundaries in narrow strips is known as

(a) Outer crop (b) Guard crop
(c) Border crop (d) All of these

96. Plants tolerant to high acidic soil known as

(a) Oxylophytes (b) Acedophytes
(c) Basilophytes (d) None of these

97. The actual area occupied by a species expressed as a percentage of total area occupied by all species in a given land area is termed as

(a) Relative dominance (b) Relative density
(c) Relative yield index (d) None of these

98. LAI represents

(a) Lower Available Index (b) Light Available Index
(c) Leaf Area Index (d) None of these

99. What is the rotation intensity of maize-potato-sugarcane-sunflower ?

(a) 100% (b) 250%
(c) 200% (d) 300%

100. The mean of plant relative yield (ratio of the per plant yields in a mixture) to that in monoculture will have a value close to unity is called :

(a) Relative yield total (b) Relative crop intensity index
(c) Relative yield index (d) None of these

101. Growing two or more generally dissimilar crop simultaneously on the same piece of land, base crop necessarily in distinct row arrangement is known as

(a) Intensive cropping (b) Inter-cropping
(c) Mixed cropping (d) None of these

102. Growing two or more crops simultaneously on the same piece of land, without any definite row arrangement is known as

(a) Intensive cropping (b) Inter-cropping
(c) Mixed cropping (d) None of these

103. Growing two or more crops simultaneously with no distinct row management is known as

(a) Mixed cropping (b) Inter-cropping
(c) Mixed inter-cropping (d) None of these

104. When rainfall is > 750 mm/annum then which type of cropping is practised ?

(a) Mono -cropping (b) Double cropping
(c) Multiple cropping (d) None of these

105. Allelopathy phenomenon is

(a) Harmful effect (b) Beneficial effect
(c) Both (a) and (b) (d) None of these

106. Annidation in time takes place in inter-cropping of

(a) Sorghum with redgram (b) Groundnut with redgram
(c) Maize with greengram (d) All the above

107. Conversion of yield of different intercrops into equivalent yield of one crop on monetary basis is known as

(a) CEY (b) RYT
(c) LER (d) MCI

108. Inter-cropping of setaria with redgram is an example of

(a) Allelopathy (b) Annidation in space
(c) Annidation in time (d) None of these

109. The yield advantage calculated on the basis of unit area and unit population is known as

(a) CEY (b) RYT
(c) LER (d) MCI

110. RYT of < 1 represents

(a) Yield reduction (b) Yield enhancement
(c) No change (d) None of these

111. A concept involving the farming of fast growing plants/trees for the purpose of providing biomass that can be used directly as fuel or converted in to other forms of energy products is known as

(a) Energy farming (b) Organic farming
(c) Diversified farming (d) Inorganic farming

112. Ratios showing food yield per unit energy inputs and are most often used in evaluating the energetics of agricultural systems is known as

(a) Energy farming (b) Organic farming
(c) Inorganic farming (d) Energy productivity ratio

113. Land left unfarmed for one or more growing seasons to kill weeds, make the soil richer is known as

(a) Fallow (b) Fallow land
(c) Fallow system (d) None of these

114. The sequence of crop years with fallow years is known as

(a) Fallow (b) Fallow land
(c) None of these (d) Fallow system

115. Extensive fallow systems are

(a) Shifting cultivation system (b) Grass fallow system
(c) Fallow system (d) None of these

116. Who proposed competition ratio ?

(a) Dalrimpal (1971) (b) Rao and Willey (1980)
(c) Monge and Associates (1978) (d) None of these

117. Who proposed cropping intensity index ?

(a) Dalrimpal (1971) (b) Rao and Willey (1980)
(c) Monge and Associates (1978) (d) None of these

118. Who proposed specific cropping intensity index ?

(a) Monge (1978) (b) Rao and Willey (1980)
(c) Monge and Associates (1978) (d) None of these

119. Who proposed cultivated land utilization index ?

(a) Choung (1973) (b) Wong, You and Strout (1978)
(c) Donald (1963) (d) Rao and Willey (1980)

120. A short duration of crop in between two main seasonal crops is known as :

(a) Cash crop (b) Inter crop
(c) Catch crop (d) None of these

121. Which of the following inter-cropping system has synergetic effect on component crop ?

(a) Autumn planted sugarcane + mustard
(b) Autumn planted sugarcane + pea
(c) Autumn planted sugarcane + rice
(d) Autumn planted sugarcane + potato

122. Which one of the following can be expressed by using equation

$$\frac{\text{Harvest diversity index}}{\text{Multiple cropping index}} \text{x}100$$

(a) Cropping intensity index (b) Multiple cropping index
(c) Simultaneous cropping index (d) None of these

123. The correct row ratio for inter-cropping of mustard with wheat is

(a) 4:1 (b) 9:1
(c) 7:1 (d) 8:1

124. Safflower can be grown as an inter-crop with

(a) Wheat (b) Potato
(c) Maize (d) Cotton

125. The example of inter-cropping in replacement series is

(a) Sugarcane + mustard (b) Sugarcane + pea
(c) Pigeonpea + urdbean (d) Potato + mustard

126. Multiple cropping increases over single cropping

(a) Soil fertility (b) Soil productivity
(c) Production of crops (d) None of these

127. Main pre-requisite for adoption of multiple cropping is

(a) Longer duration varieties (b) Soil conservation
(c) Irrigation facility (d) Weedicides

128. Crop rotation intensity is equivalent to

(a) $\frac{\text{No. of crops in a rotation}}{\text{Duration of crop rotation}} \text{x}100$

(b) $\frac{\text{No. of crops in a rotation}}{\text{No. of crops rotation}} \text{x}200$

(c) $\frac{\text{Duration of rotation}}{\text{Number of crops in roration}} \text{x } 100$

(d) $\frac{\text{Duration of crop rotation}}{\text{Number of crop rotation}} \text{x}200$

129. What is the cropping intensity of the rotation green manure - wheat ?

(a) 50% (b) 100%

(c) 200% (d) 250%

130. Rotation rice - pea - sugarcane - ratoon is of

(a) 1 year (b) 2 year

(c) 3 year (d) None of these

131. In mixed cropping

(a) Growing of two crops together

(b) Growing of two crops together in different lines

(c) Sowing of mixed seeds of two crops

(d) None of these

132. NAR, represents

(a) Net Assimilation Rate (b) Night Assessment Ratio

(c) Net Assimilation Ratio (d) None of these

133. Which one of the following cropping system is most prevalent in India ?

(a) Rice - Wheat (b) Rice - Soybean

(c) Maize - Cotton (d) None of these

134. Inter-cropping of mustard with potato is recommended in

(a) Replacement series (b) Additive series

(c) Replacement cum additive series (d) None of these

135. Example of parallel inter-cropping is

(a) Maize + soybean (b) Sugarcane + pea

(c) Potato + mustard (d) Sugarcane + mustard

136. Maize - toria - wheat - moong rotation is termed as a system of

(a) Mixed cropping (b) Relay cropping

(c) Multiple cropping (d) None of these

137. Growing plants of different heights in the same field at the same time is termed as

(a) Multiple cropping (b) Multi-storeyed cropping

(c) Relay cropping (d) None of these

138. What is *Diara* lands ?

(a) Areas flooded by sea

(b) Areas eroded by sand storms

(c) Areas on both sides of canal

(d) Areas located on either side of river which get flooded every year

139. The intensity of following one year crop rotation Maize-potato-onion is

(a) 100% (b) 150%
(c) 200% (d) 300%

140. What is inter-cropping ?

(a) Growing two or more crops simultaneously and intermingled
(b) Growing two or more crops simultaneously in definite row proportions
(c) Growing more than one crop on same land in one year
(d) None of these

141. What is the intensity of maize-potato-onion ?

(a) 100% (b) 150%
(c) 200% (d) 300%

142. Who proposed competition index ?

(a) Choung (1973) (b) Wong,You and Strout (1978)
(c) Donald (1963) (d) None of these

143. Who proposed diversity index ?

(a) Choung (1973) (b) Wong,You and Strout (1978)
(c) Donald (1963) (d) None of these

144. Who proposed aggressivity ?

(a) Choung (1973) (b) Wong,You and Strout (1978)
(c) Mc Gilchrist (1965) (d) None of these

145. Who proposed Net assimilation rate?

(a) Rao and Willey (1980) (b) Dalrimpal (1971)
(c) Evans (1982) (d) None of these

146. Who proposed leaf area index ?

(a) Choung (1973) (b) Radford (1967)
(c) Donald (1963) (d) None of these

147. A rotation of arable crops requiring annual cultivation and artificial pastures occupying field for two years or longer is called

(a) Ley farming (b) Alley cropping
(c) Silvi pastoral system (d) Strip farming

148. The practice of placing primary emphasis on the use of stored soil moisture while determining whether or not to establish a crop is known as

(a) Opportunity cropping (b) Diversified cropping
(c) Relay cropping (d) None of these

149. Crops which are grown during the rainy season is known as

(a) *Kharif* crops (b) *Perennial* crops
(c) *Zaid* crops (d) None of these

150. Crops which are grown during the winter season is known as

(a) *Kharif* crops (b) *Rabi* crops
(c) *Zaid* crops (d) Biennial crops

151. Crops which are grown during the summer season is known as

(a) *Kharif* crops (b) *Rabi* crops
(c) *Zaid* crops (d) None of these

152. A system of soil management wherein the plants are grown in permanent grass without tillage is known as

(a) Silage crop (b) Sod culture
(c) Soilage crop (d) Aquaculture

153. Forage cut green and fed to livestock while in the fresh conditions are known as

(a) Silage crop (b) Sod culture
(c) Soilage crop (d) Hydroponics

154. A crop grown in pure stand at optimum population is known as :

(a) Sole crop (b) Silage crop
(c) Strip crop (d) Inter crop

155. Growing soil covering and soil depleting crops in alternate strips running perpendicular to the slope of the land or to the direction of prevailing winds for the purpose of reducing erosion is known as

(a) Sole cropping (b) Inter-cropping
(c) Strip cropping (d) Mixed cropping

156. The crop which is grown under certain problematic conditions when no other crop can be grown as floating rice in flood prone situation is called

(a) Subsistence crop (b) Subsistence farming
(c) Substitution farming (d) Organic farming

157. Intensive fallow systems are :

(a) Jhoom cultivation system (b) Bush fallow system
(c) Fallow system (d) None of these

158. The crops are grown in parallel strips laid out across the general slope but which do not fallow the contour is known as

(a) Field strip cropping (b) Contour strip cropping
(c) Shifting cultivation (d) None of these

159. Yield of the plant parts of economic interest (economic yield) as percentage of total biological yield in terms of dry matter is called

(a) Harvest index (b) Harvest ratio
(c) Both of these (d) None of these

160. Maximum use of the land by means of frequent succession of harvested crop is known as

(a) Intensive cropping (b) Inter-cropping
(c) Mixed inter cropping (d) None of these

161. System in which crops are planted in flooded areas as the rainy season ends and water recedes is known as

(a) Recession farming (b) Diara land farming
(c) Precision farming (d) Contour farming

162. The population of a particular species expressed as a percentage of all the species in a habitat is called :

(a) Relative dominance (b) Relative density
(c) Relative yield index (d) None of these

163. In which cropping system Land Equivalent Ratio (LER) is required ?

(a) Inter-cropping system (b) Sole cropping system
(c) Mixed cropping system (d) None of these

164. Growing two or more crops simultaneously where one or more crops are planted in rows is known as

(a) Row inter-cropping (b) Row mixed cropping
(c) Strip cropping (d) Sole cropping

165. These crops which are preserved in a succulent condition by partial fermentation in a tight receptacle is known as

(a) Silage crop (b) Sod culture
(c) Soilage culture (d) Pot culture

166. Plant type or strain within a species, resulting from exposure to a particular environment is termed as

(a) Ecological amplitudes (b) Ecological pyramid
(c) Ecological niche (d) Eco-type

167. Mark the Land Equivalent Ratio is beneficial inter-cropping

(a) Zero (b) More than 1
(c) 1 (one) (d) None of these

168. Which is an example of parallel cropping ?

(a) Sugarcane + potato (b) Potato + mustard
(c) Potato + radish (d) Wheat + pea

169. What is the rotation intensity of maize - wheat + gram - moong ?

(a) 200% (b) 250%
(c) 300% (d) 400%

170. The correct row ratio for inter-cropping of mustard with gram is

(a) 5:1 (b) 6:2
(c) 9:1 (d) 8:2

171. In an inter-cropping system, if increase in the output of one crop help increase in the output of other crop, it is termed as

(a) Competitive (b) Complementary
(c) Supplementary (d) None of these

172. Certain fast growing crops which work as support to vine crop is called :

(a) Sugar crop (b) Support crop
(c) Guard crop (d) Sole crop

173. Growing of two crops in a definite row proportion on the same piece of land is known as

(a) Sole cropping (b) Inter-cropping
(c) Sole cropping (d) None of these

174. A type of agro-forestry system in which trees are planted with food crops which serve as a nursery for the young tree seedlings is known as

(a) Taungya cultivation (b) Jhoom cultivation
(c) Relay cultivation (d) Strip cultivation

175. Crops which yield in tones and grown for different market requiring heavy transport is known as

(a) Truck crops (b) Guard crops
(c) Nurse crops (d) Silage crops

176. Evaluation of inter-cropping is done by :

(a) Rotational intensity (b) Cropping intensity
(c) Land Equivalent ratio (d) None of the above

177. What is the recommended row ratio for inter-cropping of wheat + mustard?

(a) 3:1 (b) 5:1
(c) 9:1 (d) 4:1

178. The soybean crop can be grown with

(a) Rice (b) Sugarcane
(c) Maize (d) Cotton

179. Substitution of existing inefficient crop with an identified efficient crop in a zone or region is known as

(a) Subsistence farming (b) Subsistence crop
(c) Substitution farming (d) Organic farming

180. It is farming enterprise which provides food and commodities just sufficient for the farming family, and there are no surplus to tell is known as

(a) Subsistence farming (b) Subsistence crop
(c) Organic farming (d) Ley farming

181. The crops grown for the purpose of producing sugar from them is known as

(a) Sugar crop (b) Support crop
(c) Silage crop (d) Nurse crop

182. Cropping intensity is equivalent to

(a) $\frac{\text{No. of crops in a rotation}}{\text{Net available area}} \times 100$

(b) $\frac{\text{No. of crop sown}}{\text{Net available}} \times 100$

(c) $\frac{\text{Total croped area}}{\text{Duration of rotation}}$

(d) $\frac{\text{No. of crop sown}}{\text{Net available sown}}$

183. Which one of the following is the example of parallel inter-cropping ?

(a) Sugarcane + Wheat (b) Sugarcane + Mustard
(c) Sugarcane + Soybean (d) Potato + Cotton

184. Presently what is the cropping intensity of India ?

(a) 100% (b) 150%
(c) 137% (d) 160%

185. What is the cropping intensity of Paddy - Wheat - Urd - bean rotation ?

(a) 400% (b) 200%
(c) 300% (d) 500%

186. What is the rotation intensity of maize - sugarcane - ratoon ?

(a) 100% (b) 150%
(c) 300% (d) 400%

187. Growing plants of different heights in the same field at the same time is termed as

(a) Multiple cropping (b) Multi-storied cropping
(c) Strips cropping (d) Relay cropping

188. Multiple cropping is

(a) Growing two or more crops in the same field in a year
(b) Growing two or more crops in sequence on the same field in a year
(c) Growing two crops in the same field in a year
(d) All of these

189. Sequential cropping is

(a) Growing two or more crops in the same field in a year
(b) Growing two or more crops in sequence on the same field in a year
(c) Growing two crops in the same field in a year
(d) All of these

190. Double cropping is

(a) Growing two or more crops in the same field in a year
(b) Growing two or more crops in sequence on the same field in a year
(c) Growing two crops in a year in sequence
(d) All of these

191. Growing two or more crops simultaneously in the same field is known as

(a) Multiple cropping
(b) Double cropping
(c) Inter-cropping
(d) Sequential cropping

192. What is strip inter-cropping ?

(a) Growing two or more crops simultaneously with no distinct row agreement
(b) Growing two or more crops simultaneously during part of life cycle of each
(c) Growing two or more crops simultaneously with one or more crops in distinct row agreement
(d) None of these

193. What is cropping system ?

(a) Cultivation of crops in a farm
(b) Cropping patterns used on a farm
(c) Cropping systems used on a farm and its interaction with farm resources, other farm enterprises and available technology which determine their makeup.
(d) None of these

194. What is Land Equivalent Ratio ?

(a) Cultivation of crops in a farm
(b) Cropping patterns used on a farm
(c) The sum of the fractions of the yield of the intercrops, relative to their sole crop yields
(d) None of these

195. Crop rotation is

(a) Growing more than one crop at a time
(b) Growing of, crop's one after other to maintain soil fertility
(c) Growing of an associate crop in between the rows of main crop
(d) None of these

196. Agro-silvi-pastoral system is a combination of agrisilviculture and

(a) Silvi-pastoral system
(b) Agri-horti system
(c) Horti pastoral system
(d) Horti-silvi-pastoral system

197. The unit of crop produced per unit of nutrient added is called

(a) Agronomic efficiency
(b) Nutrient use efficiency
(c) Nutrient uptake
(d) None of these

198. The term 'Cropping Scheme' means

(a) Allotment of crops to the pieces of land on a farm
(b) Allotment of crop rotations to the pieces of land in a farm
(c) Planning and budgeting of the crop production in a farm
(d) None of these

199. What is "Trap crops" ?

(a) Crops grown around the main crop which are more susceptible to insect pest and diseases than the main crop
(b) Crops which are resistant to pest and diseases
(c) Crops efficient in trapping solar radiation
(d) None of these

200. A short duration crop in between two main crops is termed as

(a) Cash crop
(b) Catch crop
(c) Companion crop
(d) None of these

201. The practise of growing four crops in a year from same piece of land in such a way that one crop is sown immediately after or just before the harvest of the previous crop is known as

(a) Mono cropping
(b) Relay cropping
(c) Multiple cropping
(d) None of these

202. What are the important crop mixtures grown in Karnataka ?

(a) Jowar with ragi or linseed or bajra and tur or safflower or gram or tur and groundnut or gram and linseed or urad or horsegram or tur and horsegram or cotton and horsegram
(b) Bajra with horsegram or ragi or tur or maize or dolichos , ragi with kodra or above or rapeseed or tur or jowar and avare or italian millet or dolichos or sessamum or horsegram or cotton
(c) Cotton with navane or groundnut or kodon or chillies or tur, wheat with safflower or safflower and jowar, gram with safflower
(d) All of these

203. An appropriate combination of different farm enterprises and the means available to the farmer to raise them successfully is called

(a) Farming system (b) Cropping system
(c) Cropping pattern (d) Farming pattern

204. Multiple Cropping Index (MCI) is depicted as

(a) Percentage (b) Hectares
(c) Quintals (d) None of these

205. Cultivation of long duration crops will result in ______in cropping intensity index (CII)

(a) Increase (b) Decrease
(c) No change (d) None of these

206. The gross cropped area divided by net cultivated area is known as

(a) MCI (b) RYT
(c) LER (d) CII

207. The zone with maximum yield and spread index is known as

(a) Agro ecological zone (b) Efficient cropping zone
(c) Deficient cropping zone (d) None of these

208. From middle of October to middle of April is the season of

(a) *Kharif* (b) *Zaid, Kharif*
(c) *Rabi* (d) *Zaid, Rabi*

209. In India the period of *Zaid rabi* season is from

(a) Beginning of May to middle of October
(b) August to January
(c) Beginning of October to middle of April
(d) Beginning of February to beginning of May

210. Two years crop rotations are practiced by Indian farmers in different parts of the country

(a) Rice - pulses
(b) Jowar or maize - wheat or gram
(c) Cotton or jowar - groundnut or jowar
(d) All of these

211. Crop rotation practised by Indian farmers cover three years period is/are

(a) Wheat - maize - sugarcane (b) Sugarcane - wheat - cotton
(c) Wheat - wheat - turia (d) All of these

212. Monoculture is harmful mainly due to :

(a) Development of harmful soil organisms
(b) Reduction in soil fertility
(c) Accumulation of allelochemicals
(d) Soil erosion

213. In India, a high percentage of cropped area is covered by rice because

(a) There is adequate rainfall
(b) Large population of India is rice eater
(c) Rice is adoptable to most type of land and agro-climatic zones
(d) None of these

214. In multiple cropping

(a) Growing > 2 crops in a year
(b) Growing > 3 crops in a year
(c) Growing > 1 crops in a year
(d) None of these

215. What is the cropping intensity of the rotation chari - berseem - rice - pea ?

(a) 100%
(b) 200%
(c) 400%
(d) 500%

216. In Wheat, mustard may be grown as inter crop in the row ratio of :

(a) 6:1 (Six rows of wheat + one row of Mustard)
(b) 9:1 (Nine rows of wheat + one row of Mustard)
(c) 12:2 (Twelve rows of wheat + two rows of Mustard)
(d) None of these

217. What is relay cropping ?

(a) Growing of more than one crop on the same land in a year
(b) Growing of four crops on the same land in a year
(c) Growing of four crops on the same land in a year in such a manner that the following crop is sown before the harvest of preceding crop
(d) Growing of two crop on the same land in a year

218. Agronomic efficiency is given as

(a) kg/ kg
(b) kg/day
(c) kg/ha
(d) mg/day

219. Which of the following important crop mixtures are grown by farmers of Uttar Pradesh ?

(a) Wheat and mustard, wheat and barley, wheat and linseed, wheat and gram, wheat and lentil, barley and mustard, barley, mustard and gram, barley and gram, barley and peas
(b) Jowar and arhar, jowar arhar and urd, jowar and lobia, jowar and moong, jowar, arhar and moong, jowar and sessamun, jowar and guar, jowar, arhar and til
(c) Bajra, arhar and urd, bajra and tur, bajra, arhar and til, bajra, arhar and groundnut, bajra and groundnut, sesamum and tur, cotton and tur, maize and urd, groundnut and tur peas and mustard, groundnut and sessamum, kodon and arhar, gram and linseed
(d) All of these

220. Growing two or more crops one after the other on the same field in a year is known as

(a) Multiple cropping (b) Double cropping
(c) Sequential cropping (d) Mono cropping

221. Crop mixtures not grown by farmers of West Bengal are

(a) Jute and cowpea, jute and khesari
(b) Paddy and khesari
(c) Wheat and linseed, wheat and gram
(d) None of these

222. The area with similar soil, vegetation and population density characteristics, resulting in a similar type of cropping system is known as

(a) Agro ecological zone (b) Efficient cropping zone
(c) Deficient cropping zone (d) None of these

223. The yield index of a crop is calculated by the formula

(a) Yield of state/ yield of country x 100
(b) Average yield of state / average of country x 100
(c) Area of state/ area of country x 100
(d) None of these

224. The spread index of a crop is calculated by the formula

(a) Yield of state/ yield of country x 200
(b) Average yield of state / average yield of country x 10
(c) Area of state/ area of country x 150
(d) Average area of state / average area of country x 100

225. Inter-cropping of sorghum with redgram is an example of

(a) Allelopathy (b) Annidation in space
(c) Annidation in time (d) None of these

226. In Odisha, which of the following crop mixtures are not grown by farmers?

(a) Gram with safflower or jowar, ragi with bajra or Kodra
(b) Mustard with linseed or horsegram
(c) Gram with linseed
(d) Ragi with maize

227. Which of the following mixed crop are not important for farmers of Punjab?

(a) Jowar and sugar, bajra and moong
(b) Maize and urad, wheat and lentil
(c) Jowar and moth, bajra and cowpea
(d) All of these

228. Which of the following crop mixtures are grown by farmers of Rajasthan ?

(a) Maize with ragi or urad or cowpea or velvet bean or gaur or senji, jowar with moong, bajra with moth

(b) Wheat with gram or lentil or barley or rapeseed, barley with gram

(c) Cotton with til or senji or moth or melons, gram with toria

(d) All of these

229. Growing four crops sequentially in a year in the same field is known as

(a) Multiple cropping (b) Double cropping

(c) Quadruple cropping (d) Relay cropping

230. Generally crop rotation practices helps to

(a) Eliminate parasites which have selective hosts

(b) Lesser use of fertilizers

(c) Lesser use of pesticides

(d) More crop yield

Fill in the blanks

1. The common practice of examining inter-cropping advantage by expressing yield in monetary term is called ________ .
2. Leaf area ratio expressed as ________ .
3. Areas located on either side of river which get flooded every year is known as ______ .
4. Growing of field crops in between two widely spaced rows of fast growing trees / shrubs is known as _______ .
5. Multiple cropping index proposed by ________ .
6. Growing of crops in aberrant situations like drought and floods is called ________ .
7. Competition ratio proposed by ________ .
8. Growing three crops in a year in sequence is called _______ .
9. Growing four crops in a year in sequence is called ________ .
10. Growing two or more crops consecutively on the same field in the same year is called ________ .
11. Production of plant product of economic importance, expressed in standard units per unit area is called ________ .
12. Cropping intensity index proposed by _______ .
13. Specific cropping intensity index proposed by _______ .
14. Growing main crop in the centre surrounded by hardy and thorny crop such safflower around pea, mesta around sugarcane and sorghum around maize is called ________ .

15. Net assimilation rate proposed by ________ .
16. Leaf area index proposed by _________ .
17. The plant geometry of wheat crop is _______ .
18. The range of environmental factors over which an organisms or a physiological process can function is called __________ .
19. Crops which are grown during the rainy season is called _______ .
20. Planting of succeeding crop before harvest of standing crop is known as _______ .

True/False

1. Cultivated land utilization index proposed by Choung (1973).**(T)**
2. Competition index proposed by Mc Gilchrist (1965). **(F)**
3. Diversity index proposed by Donald (1963). **(T)**
4. Crop should be harvested at physiological maturity. **(T)**
5. Crop which are grown during the winter season is called *Zaid* crops. **(F)**
6. Crops which are grown during the summer season is called *Zaid* crops. **(T)**
7. Growing of two crops in a definite row proportion on the same piece of land is called Inter-cropping. **(T)**
8. NAR refers to Net Assimilation Rate. **(T)**
9. When sub-crops are sown to supplement the yield of the main crops, the sub-crops are known as Avenue-crops. **(F)**
10. These crop plants are grown along the farm roads and fences is called Avenue crops. **(T)**

Answers Fill in the Blanks

1. Monetary Advantage
2. $cm^{2/g}$
3. Diara lands
4. Alley Cropping
5. Dalrimpal (1971)
6. Contingency Cropping
7. Rao and Willey (1980)
8. Tripple Cropping
9. Qqudruple Cropping
10. Multiple Cropping
11. Crop productivity
12. Monge and Associates (1978)
13. Monge (1978)
14. Guard Crop
15. Dalrimpal (1963)
16. Donald (1963)
17. Extreme rectangular
18. Ecological amplitudes
19. *Kharif* Crops
20. Relay Cropping

ANSWERS

1.(c)	2.(d)	3.(a)	4.(b)	5.(b)	6.(a)	7.(c)
8.(a)	9.(c)	10.(a)	11.(d)	12.(d)	13.(a)	14.(b)
15.(c)	16.(b)	17.(c)	18.(c)	19.(a)	20.(a)	21.(d)
22.(d)	23.(a)	24.(b)	25.(a)	26.(a)	27.(a)	28.(a)
29.(a)	30.(a)	31.(d)	32.(b)	33.(b)	34.(b)	35.(a)
36.(b)	37.(c)	38.(d)	39.(c)	40.(c)	41.(b)	42.(b)
43.(c)	44.(c)	45.(a)	46.(a)	47.(a)	48.(d)	49.(a)
50.(b)	51.(d)	52.(d)	53.(a)	54.(b)	55.(c)	56.(d)
57.(c)	58.(c)	59.(c)	60.(a)	61.(d)	62.(c)	63.(c)
64.(a)	65.(c)	66.(d)	67.(a)	68.(b)	69.(c)	70.(b)
71.(b)	72.(a)	73.(a)	74.(b)	75.(a)	76.(c)	77.(c)
78.(a)	79.(a)	80.(c)	81.(d)	82.(d)	83.(a)	84.(b)
85.(b)	86.(c)	87.(a)	88.(d)	89.(a)	90.(d)	91.(a)
92.(b)	93.(c)	94.(a)	95.(d)	96.(a)	97.(a)	98.(c)
99.(c)	100.(a)	101.(b)	102.(c)	103.(c)	104.(b)	105.(a)
106.(d)	107.(a)	108.(b)	109.(b)	110.(a)	111.(a)	112.(d)
113.(b)	114.(b)	115.(a)	116.(b)	117.(c)	118.(a)	119.(a)
120.(c)	121.(d)	122.(c)	123.(b)	124.(a)	125.(d)	126.(c)
127.(c)	128.(a)	129.(b)	130.(c)	131.(c)	132.(a)	133.(a)
134.(a)	135.(a)	136.(b)	137.(b)	138.(d)	139.(d)	140.(b)
141.(d)	142.(c)	143.(c)	144.(c)	145.(b)	146.(c)	147.(a)
148.(a)	149.(a)	150.(b)	151.(c)	152.(a)	153.(c)	154.(a)
155.(c)	156.(a)	157.(b)	158.(a)	159.(a)	160.(a)	161.(c)
162.(b)	163.(a)	164.(a)	165.(a)	166.(d)	167.(b)	168.(b)
169.(d)	170.(b)	171.(b)	172.(b)	173.(b)	174.(a)	175.(a)
176.(c)	177.(c)	178.(c)	179.(c)	180.(a)	181.(a)	182.(a)
183.(c)	184.(c)	185(c)	186.(a)	187.(b)	188.(a)	189.(b)
190.(c)	191.(c)	192.(d)	193.(c)	194.(c)	195.(b)	196.(a)
197.(a)	198.(b)	199.(a)	200.(b)	201.(b)	202.(d)	203.(a)
204.(a)	205.(b)	206.(d)	207.(a)	208.(c)	209.(d)	210.(d)
211.(d)	212.(b)	213.(a)	214.(b)	215.(c)	216.(b)	217.(c)
218.(a)	219.(d)	220.(c)	221.(a)	222.(a)	223.(b)	224.(d)
225.(c)	226.(a)	227.(b)	228.(d)	229.(c)	230.(a)	

9

Weed Management

1. **Origin of *Cyperus rotundus* is _____ .**

   (a) India (b) Central America
   (c) Euresia (d) China

2. ***Amaranthus* spp. is a ________ .**

   (a) C3 plant (b) C4 plant
   (c) Both (a) and (b) (d) None of these

3. **Which is the most problematic weed present in 92 countries ?**

   (a) *Cynodon dactylon* (b) *Cyperus rotundus*
   (c) *Brecharia mutica* (d) *Argemone mexicana*

4. **Who was the first person to use the word weed in the present day of meaning ?**

   (a) Misterlisch (b) J. Tull
   (c) Molisch (d) Mohr

5. **Trade name of a combination of Atrazine + Simazine is ______ .**

   (a) Printex (b) Acetamide
   (c) Butachlor (d) Gramaxone

6. **Which one of the following is a Pseudo cereal ?**

   (a) Buck wheat (b) Dicoccum wheat
   (c) Duram wheat (d) None of these

7. **What is the trade name of propanil ?**

   (a) Stam F-34 (b) Stomp
   (c) Treflan (d) None of these

8. **What is the trade name of trifluralin ?**

   (a) Treflan (b) Stomp
   (c) Avadex (d) None of these

9. Metsulfuron methyl is a herbicide belonging to which group ?

(a) Carbamates (b) Sulfonylurea
(c) Uracil (d) None of these

10. First biologically controlled weed is

(a) *Parthenium* (b) Lantana
(c) Amaranthus (d) None of these

11. The trade name of oxadiazon is

(a) BHC (b) Bluestar
(c) Gramaxone (d) Ronstar

12. The shape of stem of *Borreria hispida* is

(a) Square (b) Round
(c) Triangular (d) All of these

13. The origin of *Echinochloa crusgalli* is

(a) Tropical America (b) Central America
(c) India (d) China

14. Dinoseb was discovered in the year ________ .

(a) 1933 (b) 1948
(c) 1928 (d) None of these

15. What is the trade name of Diquat ?

(a) Rogor (b) 2, 4 - D
(c) Reglone (d) None of these

16. Orobanche can be survive on which crop

(a) Tobacco (b) Potato
(c) Cotton (d) All of these

17. What is weed ?

(a) Unwanted plant (b) Undesirable plant
(c) Plant out of place (d) All of these

18. In India weeds account for losses dur to pests as high as

(a) 45% (b) 55%
(c) 25% (d) 35%

19. Globally weeds account for losses ______ of the total food production

(a) 11.5% (b) 16.0%
(c) 9.5% (d) 24.0%

20. Which of the following weed brought reduction in land value ?

(a) *Cyndon dactylon* (b) *Cyperus rotundus*
(c) *Agremone mexicana* (d) *Melilotus indica*

21. Which weed results in reduction in human efficiency ?

(a) *Amborsia* (b) *Parthenium*
(c) *Argemone mexicana* (d) All of these

22. The type of dormancy is found in weed seeds is

(a) Innate (b) Induced
(c) Enforced (d) All of these

23. Enforced dormancy may be due to

(a) Deeper placement of seeds (b) Genetic factor
(c) Physiological change (d) All of these

24. Innate dormancy may be due to

(a) Deeper placement of seeds (b) Genetic factor
(c) Physiological change (d) All of these

25. Induced dormancy may be due to

(a) Deeper placement of seeds (b) Genetic factor
(c) Physiological change (d) All of these

26. An ingredient which when added to a formulation enhances action of toxicant is known as

(a) Adjuvant (b) Surfactant
(c) Antidote (d) None of these

27. The amount of active ingredient given in terms of parent acid is

(a) Active ingredient (b) Acid equivalent
(c) Phytochemical (d) None of these

28. The active toxic material present in the formulation is known as

(a) Active ingredient (b) Acid equivalent
(c) Phytochemical (d) None of these

29. Which is an alternate host for hairy caterpillar?

(a) Amaranthus (b) *Crotolaria* sp.
(c) *Echinocloa* sp. (d) *Agropyron repens*

30. Which is the alternate host for stem borer ?

(a) *Melilotus alba* (b) *Crotolaria* sp.
(c) *Echinocloa* sp. (d) *Agropyron repens*

31. The dormancy of weed seeds due to presence in deeper layers is known as

(a) Induced dormancy (b) Enforced dormancy
(c) Innate dormancy (d) *In situ* dormancy

32. The dormancy of weed seeds due to genetical factors is called

(a) Induced dormancy (b) Enforced dormancy
(c) Innate dormancy (d) *In situ* dormancy

33. The induced dormancy of weed seeds is due to

(a) High temperature (b) High CO_2
(c) Water-logging (d) All of these

34. Which is a type of dormancy ?

(a) Induced (b) Enforced
(c) Innate (d) All of these

35. Which is not a type of dormancy ?

(a) Forced (b) Enforced
(c) Innate (d) Induced

36. Which of the following is not a physical method of weed control ?

(a) Cutting (b) Digging
(c) Planting (d) None of these

37. 2, 4-D was first used as a herbicide for weed control in the year

(a) 1936 (b) 1956
(c) 1944 (d) 1975

38. Which of the following is a selective herbicide ?

(a) Diquat (b) Paraquat
(c) Diallate (d) All of these

39. Which of the following is a non selective herbicide ?

(a) Glyphosate (b) Paraquat
(c) Diquat (d) All of these

40. Glyphosate belongs to which group of herbicide?

(a) Aliphatics group (b) Amide group
(c) Carbamates group (d) Thiocarbamates group

41. Propanil a postemergence herbicide is readily absorbed by the

(a) Stem (b) Root
(c) Leaf (d) Seed

42. *Phalaris minor* is a crop associated weed with

(a) Wheat (b) Sugarcane
(c) Potato (d) Pea

43. Which is the most common herbicide used in wheat ?

(a) 2,4 - D (b) Atrazine
(c) Isoproturon (d) Pendimethalin

44. Use of which of the following herbicide provides effective control of weeds in groundnut ?

(a) 2,4 - D (b) MCPB
(c) Isoproturon (d) Pendimethalin

45. Conversion of a non-phytotoxic compound into phytotoxic compound process is known as

(a) Beta oxidation (b) Oxidation
(c) Reduction (d) Alpha oxidation

46. Gesatop is the commercial name of the herbicide

(a) Simazine (b) Atrazine
(c) Fluchloralin (d) 2, 4 - D

47. Which one of the following is couch grass ?

(a) Sorghum (b) Sugarcane
(c) Potato (d) *Agropyron repens*

48. The first herbicide used is

(a) Nitric acid (b) Sodium chlorate
(c) Ammonium sulphamate (d) Copper sulphate

49. Dalapon belongs herbicide group

(a) Haloalkanoic acid (b) Phenoxyalkanoic acid
(c) Nitriles (d) None of these

50. MCPA belongs to which group of herbicide ?

(a) Haloalkanoic acid (b) Phenoxybutyric
(c) Aromatic acids (d) Nitriles

51. Gesaprin and primatol is the commercial name of the herbicide

(a) Simazine (b) Atrazine
(c) Fluchloralin (d) Butachlor

52. The herbicide group known to inhibit pantothemic acid synthesis in plants is

(a) Thiocarbamate (b) Sulfonylurea
(c) Amide (d) Acetamide

53. The phytotoxicity of monuron in cotton increases when applied with

(a) Urea (b) DAP
(c) Phorate (d) Dinosels

54. Synergistic phytotoxicity is observed on oats with phorate and

(a) Urea (b) DAP
(c) Monuron (d) Butachlor

55. **Which is the most problematic weed in deep water rice ?**

(a) *Echinocloa crusgalli* (b) Water hycinth
(c) *Cyperus rotundus* (d) *Taraxacum stricta*

56. **Weeds in pearlmillet can be controlled by spraying ________ .**

(a) Isoproturon (b) Pendimethalin
(c) Atrazine (d) 2,4 - D

57. **The control of monocot weeds in wheat crop can be done by spraying**

(a) 2,4 - D herbicide (b) Metsuffuron methyl herbicide
(c) Isoproturon herbicide (d) Atrazine

58. **In irrigated areas, the competition by weeds will be more for ______ than for water.**

(a) Light (b) Space
(c) Nutrients (d) None of these

59. **The dispersion of finely divided solid particles in the liquid is known as**

(a) Absorption (b) Suspension
(c) Sublimation (d) Fumigation

60. **All the three types, namely, enforced, innate and induced dormancies are present in which of the following weed plant ?**

(a) Wild oat (b) Amaranthus
(c) Pathenium (d) None of these

61. **Reduction in yield of maize due to weeds is less in ________ .**

(a) Harvesting stage (b) Seedling stage
(c) Knee height stage (d) Flowering stage

62. **Typha is an example of which type of aquatic weed ?**

(a) Free floating (b) Submerged
(c) Marginal (d) All of these

63. **Surfactants, based on chemical nature, are classified into**

(a) 6 group (b) 2 group
(c) 3 group (d) 5 group

64. **Based on use, surfactants are classified into**

(a) Wetting agents (b) Spreaders
(c) Penetrants (d) All of these

65. **The application of herbicides on the field before seeding or planting is termed as**

(a) Pre emergence application (b) Pre plant application
(c) Pre sowing application (d) Post sowing application

66. Chick weed is used as ________ pollution indicator in the air.

(a) NO_2 (b) SO_4
(c) CO_2 (d) SO_2

67. The herbicide Nitrogen is grouped under ______ group.

(a) Amides (b) Anilines and nitro-phenols
(c) Substituted urea (d) Acetamide

68. Flooding kills the weeds by excluding the _____ from their environment.

(a) Air (b) Water
(c) Nutrients (d) Temperature

69. A liquid dispersed in another liquid is known as

(a) Suspension (b) Emulsion
(c) Concentrate (d) Sublimation

70. *Parthenium hysterophorus* is native of ______

(a) Africa (b) India
(c) China (d) Mexico

71. Butachlor and propanil belong to which group of herbicide ?

(a) Phenoxy acid (b) Carbamates
(c) Acetamide (d) Amides

72. Alternate host of gram caterpillar is one of the following :

(a) Amaranthus (b) *Crotalaria juncea*
(c) *Agropyron repens* (d) *Argemone mexicana*

73. Skin allergy is caused by which one of the following weed ?

(a) Agropyron (b) Brush weeds
(c) *Cyperus rotundus* (d) Parthenium

74. One of the following weeds belong to the Leguminaecae family

(a) *Crotolaria juncea* (b) *Melilotus alba*
(c) *Panicum repens* (d) *Cyperus rotundus*

75. Which one of the following weedicide is recommended for wheat + mustard ?

(a) 2, 4 - D (b) Propanil
(c) Isoproturon (d) MCPA

76. Which is an example of worst perennial weed ?

(a) *Paspalum conjugatum* (b) *Argemone maxicana*
(c) *Cyperus deformis* (d) *Chenopodium album*

77. _______ is used in slash and burn cultivation to compete with *Lantana camara*

(a) *Chromolaena odorata* (b) Cascuta
(c) Orobanche (d) Striga

78. **The word allelopathy is derived from**

(a) Greek language (b) Latin language
(c) Persian language (d) French language

79. **The break down of herbicides inside the plant into nontoxic compounds is termed as**

(a) Conjugation (b) Reverse metabolism
(c) Metabolism (d) Detoxification

80. **Which herbicides is first used in India for the selective control of weed in wheat ?**

(a) Isoproturon (b) 2, 4 - D
(c) Pendimethalin (d) Butachlor

81. **2, 4- D is used for which type of weeds ?**

(a) Grassy (b) Broad leaved
(c) Both (a) and (b) (d) None of these

82. **Which toxic constituent produced by *Lantana camara* ?**

(a) Lontadens (b) Lentacil
(c) Bromocil (d) None of these

83. ***Boreria hispida* belongs to family _______ .**

(a) Palmae (b) Euphorbiaceae
(c) Verbanaceae (d) Liliaceae

84. **Which is an adsorbent ?**

(a) Activated charcoal (b) Napthalic anhydride
(c) Ammonia (d) None of these

85. **Which of the following is an adsorbent and safener ?**

(a) Activated charcoal (b) Napthalic anhydride
(c) Ammonia (d) None of these

86. **Among the chemical classification of herbicide which is an oldest group ?**

(a) Triazines (b) Amide
(c) Nitrophenols (d) None of these

87. **Dodder is a total stem parasite on**

(a) Lucern (b) Sugarcane
(c) Pea (d) Muskmelon

88. **Non-selective herbicide is _________ .**

(a) Atrazine (b) Butachlor
(c) Paraquat (d) All of these

89. Which one of the following is not narrow leaved weeds ?

(a) *Cynodon dactylon* (b) *Cyperus rotundus*
(c) *Setaria glauca* (d) *Melilotus indica*

90. Non kharif season weed is/are

(a) *Chenopodium album* (b) *Amaranthus viridisi*
(c) *Echinocloa crusgalli* (d) All of these

91. The theoretical yield of parent acid form an active ingredient is known as

(a) Active ingredient (b) RF value
(c) Acid equivalent (d) None of these

92. The weeds which grow well characteristically in acidic soils having pH between 4.5 and 6.5 is known as

(a) Acidophiles (b) Basophiles
(c) Sciophiles (d) Both (b) and (c)

93. The plants which are undesirable regardless of time and place are called

(a) Absolute weed (b) Alien weeds
(c) Relative weeds (d) Facultative weeds

94. The active component of a formulation / product is termed as

(a) Active ingredient (b) Technical ingredient
(c) Acid value (d) RR value

95. When a weed is allowed to move from place of its origin to a new area and it establishes itself there, it becomes an introduced weed in its new environment ? Such weeds are known as

(a) Alien weeds (b) Invader weeds
(c) Relative weeds (d) Absolute weeds

96. Anthrophytes are

(a) Alien weeds (b) Absolute weeds
(c) Obligate weeds (d) Facultative weeds

97. Maximum number of seeds is produced by

(a) *Echinochloa colonum* (b) *Parthenium* sp.
(c) *Utrica* sp. (d) *Orobanche cernua*

98. Maximum number of seeds is produced by

(a) *Echinochloa colonum* (b) *Parthenium* sp.
(c) *Utrica* sp. (d) *Argemone mexicana*

99. Maximum number of seeds is produced by

(a) *Amaranthus* sp. (b) *Parthenium hysterophorus*
(c) *Utrica* sp. (d) *Argemone mexicana*

100. Which one of the following weed species has maximum viability ?

(a) *Convolvulus arvensis* (b) *Meliltous indica*
(c) *Setaria glauca* (d) Both (b) and (c)

101. Wild oat exhibits which type of dormancy ?

(a) Innate dormancy (b) Induced dormancy
(c) Enforced dormancy (d) All of these

102. Which of the following factors does not affect dormancy ?

(a) Temperature (b) Light
(c) Soil fertility (d) None of these

103. Dormancy is due to

(a) Seed coat (b) Immute embryo
(c) Both (a) and (b) (d) None of these

104. Which is an corn perennial weed ?

(a) *Phalaris minor* (b) *Cyperus rotundus*
(c) *Timothy* (d) None of these

105. Which is an creeping perennial weed ?

(a) *Cynodon dactylon* (b) *Cyperus rotundus*
(c) *Convolvulus arvensis* (d) All of these

106. Which one of the following is a dryland weed ?

(a) *Phalaris minor* (b) *Cyperus rotundus*
(c) *Cynotis cucullata* (d) All of these

107. Which is a wet land weed ?

(a) *Phalaris minor* (b) *Cyperus rotundus*
(c) *Eclipta prostrata* (d) None of these

108. Which is a acidophile weed ?

(a) *Phalaris minor* (b) *Argemone mexicana*
(c) *Rumex acetosella* (d) None of these

109. *Loranthus longifoloius* weed is

(a) Total stem parasitic (b) Partial root parasitic
(c) Total root parasitic (d) Partial stem parasitic

110. *Fimbrystylis miliaceae* weed is

(a) Total stem parasitic (b) Partial root parasitic
(c) Total root parasitic (d) None of these

111. In the wheat field commonly found annual weed is

(a) *Allagallis arvellsis* (b) *Echinochola crusgalli*
(c) *Amaranthus minor* (d) *Phalaris minor*

112. The herbicides containing carbon and hydrogen in their molecule are known as

(a) Arsenic (b) Inorganic herbicide
(c) Organic herbicide (d) None of these

113. In the following non selective herbicide is

(a) Alachlor (b) Butchlor
(c) Paraquat (d) All of these

114. Which one of the following is not narrow leaved weeds ?

(a) *Cynodon dactylon* (b) *Cyperus rotundus*
(c) *Setaria glacua* (d) *Melilotus indica*

115. Which chemical is used for suicidal germination of striga seeds ?

(a) Ethylene (b) Bromine
(c) Methylene (d) Chlorine

116. Which is not a weed control measure ?

(a) Chemical (b) Cultural
(c) Physical (d) None of these

117. Which is a weed control measure ?

(a) Biological (b) Dormancy
(c) Allelopathy (d) None of these

118. Which is not a cultural weed control method ?

(a) Fertilizer application (b) Irrigation
(c) Herbicide application (d) None of these

119. Which is not a physical method of weed control ?

(a) Cutting (b) Digging
(c) Planting (d) None of these

120. Which weeds most suitable for biological control ?

(a) Faculative weeds (b) Obligate weeds
(c) Introduced weeds (d) Associated weeds

121. Dicamba belongs to which group of herbicide ?

(a) Nitriles (b) Phenoxybutyric acid
(c) Aromatic acids (d) None of these

122. Bromoxynil belongs to which group of herbicide ?

(a) Haloalkanoic acids (b) Phenoxybutyric
(c) Nitriles (d) None of these

123. Which one of the following is a post emergence grass killer used in dicotyledonous crops ?

(a) Propanil (b) Dalapon
(c) Fluazifop-butyl (d) Bromoxynil

124. Which of the following is a total weed killer used in non cropped areas ?

(a) Chlorofenac (b) Dalapon
(c) Propanil (d) Bromoxynil

125. Plant growing where it is not desired is termed as

(a) Off types (b) Weeds
(c) Crop plants (d) None of these

126. Plant whose economic value has not been discovered is called

(a) Off types (b) Weeds
(c) Crop plants (d) All of these

127. Which is a non-selective systemic herbicide ?

(a) Bromoxynil (b) Dalapon
(c) Glyphosate (d) None of these

128. Coupling of intact herbicide molecules with plant cell is termed as

(a) Conjugation (b) Addehision
(c) Sticking (d) Mutation

129. A substance that is added to herbicide to increase its efficiency is known as

(a) Adjuvant (b) Surfactant
(c) Sticker (d) Fumigant

130. Paraquat is available in the form of _______ .

(a) Wettable powder (b) Soluble concentrate
(c) Emulsifiable concentrate (d) Fumigants

131. Which one of the following was introduced in the country to eradicate prickly pear (*Opuntia dilleni*) weed ?

(a) *Chrysomella* spp. (b) *Cyperus rotundus*
(c) *Brecharia mutica* (d) *Setaria glauca*

132. Which one of the following microbial agents is being commercially exploited as biocontrol agent ?

(a) *Penicillium notatum* (b) *Bacillus subtilis*
(c) *Trichoderma viridiae* (d) *Sclerotium rolfsii*

133. Which plant can grow well under acidic soil conditions ?

(a) Acedophiles (b) Halophytes
(c) Sciophytes (d) None of these

134. Which one of the following causes more wastage of herbicide by drift ?

(a) Low volume sprayer (b) Ultra- low volume sprayer
(c) Hand sprayer (d) Both (b) and (c)

135. In the process of functional allelopathy

(a) Toxic substances are released as such from the plant
(b) A precursor is released which is converted into active substances by some micro-organisms
(c) Release of nitrogen from nodule of legume take place
(d) All of these

136. The herbicides containing carbon and hydrogen in their molecule are known as

(a) Inorganic herbicide (b) Acid
(c) Organic herbicide (d) None of these

137. Which is a basophile weed ?

(a) *Phalaris minor* (b) *Cyperus rotundus*
(c) *Taraxacum stricta* (d) None of these

138. Which is a neutrophile weed ?

(a) *Taraxacum stricta* (b) *Rumex acetosella*
(c) *Achylapha indica* (d) All of these

139. Which weeds grow in acid soil ?

(a) Neutrophile weed (b) Acidophile weed
(c) Basophile weed (d) All of these

140. Which weed grows in neutral soil ?

(a) Neutrophile weed (b) Acidophile weed
(c) Basophile weed (d) All of these

141. Which weeds completing their life cycle in a year are ______ .

(a) Annual weeds (b) Biennial weeds
(c) Perennial weeds (d) None of these

142. Weeds that grow in on season and completing their life cycle in the following season/ year are known as

(a) Annual weeds (b) Biennial weeds
(c) Perennial weeds (d) None of these

143. Weeds surviving for more than two years are known as

(a) Annual weeds (b) Biennial weeds
(c) Perennial weeds (d) None of these

144. Which is an annual weed ?

(a) *Bidens pilosa* (b) *Chenopodium album*
(c) *Phalaris minor* (d) All of these

145. What is the chemical name of Dalapon ?

(a) 2, 2, Dichloropropionic acid (b) 2, 4, Dichloropropionic acid
(c) 2, 6, Dichloropropionic acid (d) Dichlorodiphenyltrichloro ethane

146. Striga is a partial root parasite on

(a) Mango (b) Sugarcane
(c) Cotton (d) Sorghum

147. Orobanche is a total root parasite on

(a) Sugarcane (b) Tobacco
(c) Sorghum (d) Pea

148. *Cassytha filiformis* is a partial stem parasite on

(a) Guava (b) Orange
(c) Banana (d) Cotton

149. *Loranthus longifolius* is a partial stem parasite on

(a) Pea (b) Jackfruit
(c) Mango (d) Tomato

150. The weeds that belong to the family ______ are known as sedges.

(a) Cyperaceae (b) Malvaceae
(c) Leguminaceae (d) Gramineae

151. Which of the following herbicides is used for chemical tillage ?

(a) 2, 4 -D (b) Grammaxone
(c) MCPA (d) Isoproturon

152. Which one of the following weedicides is available in wettable powder form ?

(a) 2, 4 -D (b) Propanil
(c) Isoproturon (d) MCPA

153. Which weed grows in saline and alkaline soil ?

(a) Neutrophile weed (b) Acidophile weed
(c) Basophile weed (d) All of these

154. Which is a monocot weed ?

(a) *Commelina benghalensis* (b) *Convolvulus arvensis*
(c) *Crotolaria verucosa* (d) None of these

155. Which is a dicot weed ?

(a) *Commelina benghalensis* (b) *Setaria gluca*
(c) *Crotolaria verucosa* (d) None of these

156. Which is a grassy weed ?

(a) *Commelina benghalensis* (b) *Setaria gluca*
(c) *Crotolaria verucosa* (d) All of these

157. Which is a broadleaf weed ?

(a) *Commelina benghalensis* (b) *Setaria gluca*
(c) *Cyperus rotundus* (d) All of these

158. Which is an example of sedges weed ?

(a) *Commelina benghalensis* (b) *Cyperus rotundus*
(c) *Filmbrystylis miliaceae* (d) Both (b) and (c)

159. Which is an example of poisinous weed ?

(a) *Commelina benghalensis* (b) *Cyperus rotundus*
(c) *Filmbrystylis miliaceae* (d) *Dautra stamonium*

160. Which is an example of parasitic weed ?

(a) *Orobanche minor* (b) *Striga hemonthica*
(c) *Cuscutta compostris* (d) All of these

161. Which is a biennial weed ?

(a) *Phalaris minor* (b) *Bidmes pilosa*
(c) *Alternenthera pungens* (d) *Convolvulus arvensis*

162. Which is a perennial weed ?

(a) *Sonchus arvensis* (b) *Cyperus rotundus*
(c) *Convolvulus arvensis* (d) All of these

163. Which is an simple perennial weed ?

(a) *Phalaris minor* (b) *Cyperus rotundus*
(c) *Sonchus arvensis* (d) All of these

164. Which is an bolbous perennial weed ?

(a) *Phalaris minor* (b) *Cyperus rotundus*
(c) *Allium* sp. (d) None of these

165. Which one of the following is not kharif season weed ?

(a) *Chenopodium album* (b) *Amaranthus viridis*
(c) *Echinocloa crusgalli* (d) *Commenia benghalensis*

166. In the following aquatic weed is

(a) *Salvinia* (b) *Amaranthus viridis*
(c) *Echinocloa crusgalli* (d) *Commenia benghalensis*

167. Which one of the following is a submersed weed ?

(a) *Cypersus rotundus* (b) *Utricularia flexosus*
(c) *Pistia stratiotes* (d) None of these

168. Which one of the following is not a floating weed ?

(a) *Nymphea pupseems* (b) *Utricularia flexosus*
(c) *Pistia stratiotes* (d) None of these

169. Which one of the following is a floating weed ?

(a) *Salvinia* (b) *Nymphea pupseems*
(c) *Pistia stratiotes* (d) All of these

170. *Cuscutta compestris* is

(a) Total stem parasitic weed (b) Partial root parasitic weed
(c) Total root parasitic weed (d) All of these

171. *Striga hemonthica* is

(a) Total stem parasitic weed (b) Partial root parasitic weed
(c) Total root parasitic weed (d) All of these

172. Orobanche minor is

(a) Total stem parasitic weed (b) Partial root parasitic weed
(c) Total root parasitic weed (d) All of these

173. Which is partial stem parasitic weed ?

(a) *Orobanche minor* (b) *Striga hemonthica*
(c) *Cuscuta compostris* (d) *Cassytha filiformis*

174. The weed which is not native of a country is known as ______ .

(a) Faculative weed (b) Alien weed
(c) Obligate weed (d) Satellite weed

175. Which of the following weed is not a perennial weed ?

(a) Bermuda grass (b) Nut grass
(c) Johnson grass (d) Canary grass

176. Which is a alien weed?

(a) *Cynodon dactylon* (b) *Cyperus rotendus*
(c) *Phalaris minor* (d) *Maxican poppy*

177. Which is a post emergence contact herbicide used in

(a) Diallale (b) Dalapon
(c) Fluazifop-butyl (d) Propanil

178. Which one of the following herbicide used for control of wild oat?

(a) Di-allate (b) Dalapon
(c) Propanil (d) None of these

179. Which is total root parasitic weed ?

(a) *Orobanche minor* (b) *Striga hemonthica*
(c) *Cuscutta compostris* (d) All of these

180. Which is partial root parasitic weed ?

(a) *Orobanche minor* (b) *Striga hemonthica*
(c) *Cuscuta compestris* (d) All of these

181. Which is total stem parasitic weed ?

(a) *Orobanche minor* (b) *Striga hemonthica*
(c) *Cuscuta compestris* (d) All of these

182. Which is an objectionable weed in lucern ?

(a) Amaranthus (b) Dodder
(c) *Echinocloa* sp. (d) *Agropyron repens*

183. Which is an objectionable weed in methi ?

(a) *Agropyron repens* (b) *Melilotus alba*
(c) *Echinocloa* sp. (d) None of these

184. The passage of toxicity from one plant to another is known as

(a) Teletoxicity (b) Phytotoxicity
(c) Xenotoxicity (d) Relative toxicity

185. Which is an objectionable weed in Barseem ?

(a) Amaranthus (b) *Cichorium intybus*
(c) *Echinocloa* sp. (d) *Agropyron repens*

186. Which is an objectionable weed in mustard ?

(a) Amaranthus (b) *Argemone maxicana*
(c) *Echinocloa* sp. (d) *Agropyron repens*

187. Which is an objectionable weed in wheat ?

(a) Amaranthus (b) *Convolvulus arvensis*
(c) *Echinocloa* sp. (d) *Agropyron repens*

188. Which is the alternate host for gram caterpillar ?

(a) Amaranthus (b) *Crotolaria* sp.
(c) *Echinocloa* sp. (d) *Agropyron repens*

189. Which is the alternate host for black rust ?

(a) Amaranthus (b) *Crotolaria* sp.
(c) *Echinocloa* sp. (d) *Agropyron repens*

190. Which is the alternate host for ergot ?

(a) *Crotolaria* sp. (b) *Cenchrus ciliaris*
(c) *Sacharum spontaneum* (d) None of these

191. Which is the alternate host for downey mildew ?

(a) *Crotolaria* sp. (b) *Cenchrus ciliaris*
(c) *Sacharum spontaneum* (d) *Agropyron repens*

192. Dermatitis is caused by

(a) Brush weed (b) Parthenium
(c) *Utrica* sp. (d) None of these

193. Hay fever (Allergic rhinitis) is caused by

(a) Ambrosia (b) Brush weed
(c) *Utrica* sp. (d) None of these

194. Itching and Inflammation is caused by

(a) Brush weed (b) Parthenium
(c) *Utrica* sp. (d) None of these

195. Which of the following is responsible for African sleeping sickness ?

(a) Ambrosia (b) Parthenium
(c) *Utrica* sp. (d) Brush weed

196. Which is an example of *Cyperus rotundus* ?

(a) Annual weed (b) Perennial weed
(c) Biennial weed (d) None of these

197. Which one of the following weedicideis recommended for wheat + mustard crop ?

(a) 2,4 -D (b) MCPA
(c) Isoproturon (d) MCPB

198. Which one of the following weedicides is available in wettable power form:

(a) MCPA (b) 2,4 -D
(c) Isoproturon (d) MCPB

199. Which one of the following herbicides is used for chemical tillage ?

(a) Isoproturon (b) Grammaxone
(c) MCPA (d) 2,4 - D

200. Which one of the following weeds belong to the family Leguminaecae ?

(a) *Argemane mexicana* (b) *Melilotus alba*
(c) *Pancium repens* (d) *Cyperus deformis*

201. Metolachlor belongs to herbicide group

(a) Aliphatics group (b) Amide group
(c) Carbamates group (d) Thiocarbamates group

202. Chlorpropham belongs to herbicide group

(a) Aliphatics group (b) Amide group
(c) Carbamates group (d) Thiocarbamates group

203. Diallete belongs to which group of herbicide ?

(a) Aliphatics group (b) Amide group
(c) Carbamates group (d) Thiocarbamates group

204. Isoproturon belongs to herbicide group

(a) Substituted Urea (b) Amide
(c) Carbamates (d) Thiocarbamates

205. The weeds that belong to the family ________ are known as sedges

(a) Liliaceae (b) Papaveraceae
(c) Euphorbiaceae (d) Cyperaceae

206. The critical period of weed competition for transplanted rice is _____ after transplanting

(a) 30-50 days (b) 10-20 days
(c) 25-55 days (d) 40-65 days

207. Chemical name for herbicide 'Machete' is :

(a) Paraquat (b) Diquat
(c) Alachlor (d) Butachlor

208. Glyphosate is a _____ .

(a) Selective herbicide (b) Pre-emergence herbicide
(c) Non - selective herbicide (d) Preplanting herbicide

209. Orobanche is a parasite on which of the following crop ?

(a) Potato (b) Sugarcane
(c) Safflower (d) Tobacco

210. Growth of parthenium weed can be controlled by which of the following weed ?

(a) *Cassia serecia* (b) *Argemone mexicana*
(c) *Cynodon dactylon* (d) *Cyperus deformis*

211. Butachlor and propanil belong to which group of herbicide ?

(a) Ureas (b) Carbamates
(c) Aliphatic (d) Amides

212. Which of the following under the trade name Collego is normally applied to rice and soybean

(a) Bioherbicide (b) Chemical herbicide
(c) Mycoherbicide (d) None of these

213. Which one of the following is the alternate host of gram caterpillar ?

(a) *Amaranthus* (b) *Crotalaria juncea*
(c) *Agropyron repens* (d) None of these

214. Which one of the following caused skin allergy ?

(a) *Amaranthus* (b) Brush weeds
(c) *Crotalaria* (d) *Parthenium*

215. Which one of the following weed comes in the category of sedges ?

(a) *Crotalaria juncea* (b) *Eclipta alba*
(c) *Cyperus rotundus* (d) *Argemone mexicana*

216. Herbicide is a chemical which used to

(a) Destroy weeds (b) Destroy crops
(c) Destroy weeds and crops both (d) Destroy pests

217. Weed competition in rice is more severe in ________

(a) Direct seeded crop (b) Late sown crop
(c) Early sown crop (d) None of these

218. The critical period of weed control in upland rice varies between ________ after sowing.

(a) 15-30 days (b) 10-15 days
(c) 35-40 days (d) 5-10 days

219. The herbicide which become active only when under go - 3-Oxidation is

(a) 2,4 - D (b) Paraquat
(c) MCPA (d) 2,4 - D B

220. Which is an example of *Cyperus rotundus* ?

(a) Grass (b) Sedge
(c) Broad leaf weed (d) None of these

221. _______ weed used against snake bite

(a) Piwala dhotra (b) Dudhi
(c) Tampa (d) Ekdanli

222. The weed with maximum number of seeds per plant is

(a) *Utrica* sp. (b) Parthenium
(c) *Lambs quarters* (d) *Cyperus rotundus*

223. Which is the world's most problematic weed ?

(a) Ambrosia
(b) Parthenium
(c) *Utrica* sp.
(d) *Cyperus rotundus*

224. The effect of one plant on germination and growth of other plant in their vicinity is

(a) Allelopathy
(b) Competition
(c) Critical period
(d) None of these

225. The shortest period during period which weeding results in maximum economic return is called

(a) Critical period of weed competition
(b) Critical period of crop competition
(c) Critical period of weed association
(d) Critical period of crop association

226. Which one of the following herbicides is a selective herbicide ?

(a) Atrazine
(b) Paraquat
(c) Glyphosate
(d) None of these

227. 2,4-D is a

(a) Selective herbicide
(b) Non selective herbicide
(c) Contact herbicide
(d) Selective and systemic herbicide

228. Most preferred herbicide for maize under dry land conditions

(a) Atrazine
(b) Paraquat
(c) Propanil
(d) Butachlor

229. Which of the following recommended for the control of weeds in blackgram?

(a) Atrazine 0.5 kg/kg
(b) 2, 4 - D 0.75 kg/ha
(c) Fluchloralin 0.75 kg/ha
(d) None of these

230. Which stages of a crop prone to weed competition ?

(a) Germination to seedling stage
(b) Vegetative stage
(c) Reproductive stage
(d) None of these

231. Atrazine is a selective herbicide for controlling weeds in which of the crop ?

(a) Maize
(b) Wheat
(c) Potato
(d) Rice

232. The complementary use of resources by two or more plants species is known as

(a) Annidation
(b) Allelopathy
(c) Antagonism
(d) None of these

233. Metabolic process in which organic acids are shortened by two carbon increments is

(a) Beta oxidation
(b) Reduction
(c) Oxidation
(d) Sublimation

234. Conversion of 2,4 - DB to 2,4 - D in some plants is an example of

(a) Beta oxidation
(b) Reduction
(c) Oxidation
(d) Neutralization

235. Enzyme responsible for conversion of 2,4 - DB to 2,4 - D in some plants is

(a) Beta oxidase
(b) Beta reductase
(c) Nitrate reductase
(d) Isomerase

236. 2,4 - DB can be used in pulses but 2, 4 - D can not be used because of

(a) Absence of β oxidase
(b) Absence of β reductase
(c) Presence of β oxidase
(d) Presence of nitrate reductase

237. Which one of the following weed introduced from some other part of the world ?

(a) *Cynoodon dactylon*
(b) *Cyperus rotendus*
(c) *Phalaris minor*
(d) Water hyacinth

238. Which is an alien weed ?

(a) *Cynoodon dactylon*
(b) *Cyperus rotendus*
(c) *Phalaris minor*
(d) Maxican poppy

239. Which is a post emergence contact herbicide used in rice ?

(a) Di-allate
(b) Dalapon
(c) Fluazifop-butyl
(d) Propanil

240. Which one of the following herbicide used for control of wild oat ?

(a) Di-allate
(b) Dalapon
(c) Propanil
(d) None of these

241. Which herbicide is used for control of weeds in groundnut?

(a) Simazine
(b) 2, 4 -D
(c) Fluchloralin
(d) Paraquat

242. *Avena fatua* weed is associated with ________ .

(a) Paddy
(b) Maize
(c) Cotton
(d) Wheat

243. _______ is a parasitic weed associated with sorghum .

(a) Hariyali
(b) Lavala
(c) Striga
(d) Ekdanli

Fill in the blanks

1. Water hyacinth weed can be grouped under ________ .
2. *Phabis minor* weed associated with _______ crop .
3. 2, 4-D is used to control ______ type of weeds.
4. Striga is a parasitic weed observed on ______ .
5. *Xanthium strumarium* [Gokharus] weed is mostly dispersed by ______ .
6. The weed which is not a native of country is known as _________ .
7. Dermatitis (skin allergy) caused by _______ .
8. The chemicals that prevent the growth of green plant when present in the soil considered as a ________ .
9. Study of stage in the weed species shift is known as ________ .
10. Most common herbicide used in maize is _________ .
11. Orabanche is a weed on ________ .
12. Urea herbicides under go conjugation with _______.
13. Propanil, a post-emergence herbicide is readily absorbed by _________ .
14. *Imperata cylindrica* is associated with ________ .
15. Dodder is a serious weed of _______ .
16. A weed that spread Asthma is ________ .
17. _________ is the first person to use the definition of weed.
18. _______ is the first herbicide produced in the world.
19. 2, 4 D is a _______ herbicide.
20. __________ is the world worst weed.

True/ False

1. Weedicide can't be dust sprayed because of drift hazard. **(T)**
2. Doob and Kans weeds are examples for dicot. **(F)**
3. ABG 500 is a bio-herbicide used against Abutilon. **(T)**
4. Soyabean is a indicator plant for the bioassay of Atrazine. **(T)**
5. Bioassay a testing method using organism to determine the presence of a herbicide effect. **(T)**
6. "CASST" is a bio-herbicide used to control *Abutilon indicum*. **(F)**
7. Adjuvants are chemical used to improve the herbicidal effect. **(T)**
8. Facultative weed means grow both wild and in association with crops. **(T)**
9. "Indian journal of weed science" is published by Tropical Weed Research. **(T)**

10. The time needed for half the pesticide to disappear from the soil is known as half time. **(F)**
11. Butachlor herbicide is applied in paddy crop. **(T)**
12. Isoproturon is a fungicide. **(F)**
13. Weed control efficiency formula has been suggested by Patil and Patil (1983). **(T)**
14. Use of Tok E-25 provides effective control of weeds in groundnut. **(T)**
15. Many weeds in India originated in some other part of the word is called alien weeds. **(T)**
16. The noxius weeds whose seeds are difficult to separate, once mixed with crop stand is called noxicious weeds. **(F)**
17. Scientific name of the weed "Hirankhuri" is *Convolvulus arvensis.* **(T)**
18. Non-selective herbicide kill all the plants. **(T)**
19. Blind tillage is normally practised to control weeds in row planting crop. **(T)**
20. Relative weeds are plants of other in the crop field. **(F)**

Answers Fill in the Blanks

1. Aquatic
2. Wheat
3. Broad leaves weeds
4. Sorghum
5. Animals
6. Alien weed
7. *Parthenium hysterophorus* and *Ambrosia* sp.
8. Soil Sterilant
9. Weed Population Dynamics
10. Simazine
11. Tobacco
12. Proteins
13. Leaf
14. Tea
15. Lucerne
16. Parthenium
17. Jethro Tull
18. 2,4 - D
19. Selective and systemic
20. *Cyperus*

ANSWERS

1.(c)	2.(b)	3.(b)	4.(b)	5.(a)	6.(a)	7.(a)
8.(a)	9.(b)	10.(b)	11.(d)	12.(a)	13.(a)	14.(a)
15.(c)	16.(a)	17.(d)	18.(a)	19.(a)	20.(b)	21.(d)
22.(d)	23.(a)	24.(b)	25.(c)	26.(a)	27.(b)	28.(a)
29.(b)	30.(c)	31.(b)	32.(c)	33.(d)	34.(d)	35.(a)
36.(c)	37.(c)	38.(c)	39.(d)	40.(a)	41.(c)	42.(a)
43.(a)	44.(d)	45.(a)	46.(a)	47.(d)	48.(d)	49.(a)
50.(b)	51.(b)	52.(b)	53.(c)	54.(c)	55.(c)	56.(c)
57.(c)	58.(c)	59.(b)	60.(a)	61.(a)	62.(c)	63.(c)
64.(d)	65.(b)	66.(a)	67.(b)	68.(a)	69.(b)	70.(d)
71.(d)	72.(a)	73.(d)	74.(b)	75.(c)	76.(a)	77.(a)
78.(a)	79.(b)	80.(b)	81.(b)	82.(a)	83.(c)	84.(a)
85.(b)	86.(c)	87.(a)	88.(c)	89.(d)	90.(a)	91.(c)
92.(a)	93.(a)	94.(a)	95.(a)	96.(a)	97.(d)	98.(a)
99.(a)	100.(a)	101.(d)	102.(c)	103.(c)	104.(c)	105.(d)
106.(c)	107.(c)	108.(c)	109.(d)	110.(d)	111.(d)	112.(c)
113.(a)	114.(d)	115.(a)	116.(d)	117.(a)	118.(c)	119.(c)
120.(c)	121.(c)	122.(c)	123.(c)	124.(a)	125.(b)	126.(b)
127.(c)	128.(a)	129.(a)	130.(b)	131.(a)	132.(c)	133.(a)
134.(a)	135.(b)	136.(c)	137.(c)	138.(c)	139.(b)	140.(a)
141.(a)	142.(b)	143.(c)	144.(d)	145.(a)	146.(d)	147.(b)
148.(b)	149.(c)	150.(a)	151.(b)	152.(c)	153.(c)	154.(a)
155.(c)	156.(b)	157.(a)	158.(d)	159.(d)	160.(d)	161.(c)
162.(d)	163.(c)	164.(c)	165.(a)	166.(a)	167.(b)	168.(b)
169.(d)	170.(a)	171.(a)	172.(c)	173.(d)	174.(b)	175.(d)
176.(d)	177.(d)	178.(a)	179.(a)	180.(b)	181.(c)	182.(b)
183.(b)	184.(a)	185(b)	186.(b)	187.(b)	188.(a)	189.(d)
190.(b)	191.(c)	192.(b)	193.(a)	194.(c)	195.(d)	196.(b)
197.(c)	198.(c)	199.(b)	200.(b)	201.(b)	202.(c)	203.(d)
204.(a)	205.(d)	206.(a)	207.(d)	208.(c)	209.(d)	210.(a)
211.(d)	212.(c)	213.(a)	214.(a)	215.(c)	216.(a)	217.(a)
218.(a)	219.(d)	220.(b)	221.(c)	222.(c)	223.(d)	224.(a)
225.(a)	226.(a)	227.(c)	228.(a)	229.(c)	230.(a)	231.(a)
232.(a)	233.(a)	234.(a)	235.(a)	236.(a)	237.(d)	238.(d)
239.(d)	240.(a)	241.(c)	242.(d)	243.(c)		

10

Dryland Management

1. **The broad beds and furrow method is generally suitable for**
   (a) Black soils
   (b) Sandy soils
   (c) Loamy soils
   (d) Sandy loam soils
2. **Evaporation from soil surface can be restricted**
   (a) by reducing external evaporativity
   (b) by reducing energy supply to evaporating surface
   (c) by using growth retardants
   (d) All of these
3. **Moisture stress in soils and plants during crop growth due to imbalance between soil moisture and evapotranspiration of a crop is____**
   (a) Agricultural drought
   (b) Hydrological drought
   (c) Meteorological drought
   (d) Physiological drought
4. **Which one be called as dryland ?**
   (a) when rainfall received is less than 1000 mm
   (b) when rainfall received is less than 400 mm
   (c) when rainfall received is less than 750 mm
   (d) None of these
5. **Transpiration can be restricted by**
   (a) Use of anti-transpirants
   (b) Increasing leaf reflectance
   (c) Use of growth retardants
   (d) All of these
6. **The zone delineated by Thornthwaite's moisture indices from more than 100 is:**
   (a) Semi-arid
   (b) Humid
   (c) Arid
   (d) Extreme Humid

7. **Thornthwaite's moisture index can be calculated by**

(a) $MI = \frac{P - PE}{PE} \times 100$

(b) $MI = \frac{P + PE}{PE} \times 100$

(c) $MI = \frac{P - PE}{P} \times 100$

(d) $MI \frac{P - PE}{PE}$

8. **A dry soil layer at the surface may be created *in-situ* by loosening the surface soil and hastening the drying of the upper soil is known as**

(a) Soil mulch (b) Straw mulch
(c) Stuble mulch (d) None of these

9. **Mulching with crop residue like straw, husk dry leaves, etc. helps to reduce surface runoff is known as**

(a) Stuble mulch (b) Straw mulch
(c) Plastic mulch (d) All of these

10. **Mulching with costly plastic films and paper is practices to reduce moisture losses from soil surface is known as**

(a) Stuble mulch (b) Straw mulch
(c) Plastic mulch (d) None of these

11. **The fraction of total precipitation which forms a part of the crop consumptive use is known as**

(a) Effective rainfall (b) Effective irrigation
(c) Effective climate (d) Erodibility

12. **Controlled and uniform application of water to crop land in required amount at required time, with minimum cost to produce optimum yields without waste of water and adverse effect on soil in form of soil salinity and water logging problems is called**

(a) Effective rainfall (b) Effective irrigation
(c) Effective climate (d) Erodibility

13. **A small body of water retained behind a small dam or held in a pit dug in the ground by water harvesting techniques is called:**

(a) Farm pond (b) Gully plug
(c) Lakes (d) Rivers

14. **Leaf and canopy temperature can be measured by:**

(a) Infra red thermometer (b) Barometer
(c) Lysimeter (d) None of these

15. **Organic substances which in minute amounts may participate in the control of growth processes is known as**

(a) Growth regulator (b) Growth retardant
(c) Growth stimulant (d) None of these

16. **The International Centre for Agricultural Research in Dry Areas (ICARDA) was established in the year**

    (a) 1987 (b) 1977
    (c) 1967 (d) 1980

17. **The ICRISAT was established in the year**

    (a) 1987 (b) 1972
    (c) 1967 (d) 1982

18. **The CRIDA was established in the year**

    (a) 1987 (b) 1982
    (c) 1967 (d) 1992

19. **Where the dryland areas can be designated**

    (a) annual rainfall is equal to potential evapotranspiration
    (b) potential evapotranspiration is more than annual rainfall
    (c) potential evapotranspiration is equal to annual rainfall
    (d) None to these

20. **In stress condition, which amino acid accumulated in crop plants**

    (a) Tryptophan (b) Isoleucine
    (c) Proline (d) Phenyl alanine

21. **Mulching is generally practiced in high value crops is:**

    (a) Stuble mulch (b) Straw mulch
    (c) Plastic mulch (d) All of these

22. **Which plastic is more effective in mulching ?**

    (a) Black plastic (b) White plastic
    (c) Transparent plastic (d) All of these

23. **Plastic mulching is also act as:**

    (a) Reduces weed infestation (b) Increases seed germination
    (c) Both (a) and (b) (d) None of these

24. **The amount of water needed to return the soil moisture status to the field capacity is known as**

    (a) Soil moisture deficit (b) Soil moisture stress
    (c) Soil moisture tension (d) None of these

25. **It is the sum of soil moisture tension and the osmotic pressure of the soil solution is called:**

    (a) Soil moisture deficit (b) Soil moisture stress
    (c) Soil moisture tension (d) None of these

26. A growth regulating chemical produced by a plant of a synthetic one with similar activity even at very low concentration is known as

(a) Growth regulator (b) Growth retardant
(c) Growth promoter (d) None of these

27. Chemical such as auxins, gibberellins, kinins etc. are the example of:

(a) Growth regulator hormones (b) Growth retardant hormones
(c) Growth substances (d) None of these

28. Reshaping of the field surface to a desired elevation and slope is known as

(a) Land forming (b) Land grading
(c) Land levelling (d) None of these

29. The reshaping of the land surface to facilitate a more uniform application of irrigation water or for preventing soil erosion is known as

(a) Land forming (b) Land grading
(c) Land leveling (d) None of these

30. Usually crop face little or no moisture stress during their life cycle in:

(a) Dry farming (b) Dry season
(c) Drought year (d) Rainfed farming

31. Rainfed farming areas are characterized by a growing season of

(a) <75 days (b) .75-120 days
(c) >120 days (d) 50-70 days

32. In India Dryland constitutes ________ of total cultivated area

(a) 72% (b) 88%
(c) 68% (d) 48%

33. Dryland contributes _______ of total food production in India

(a) 72% (b) 84%
(c) 68% (d) 42%

34. In rainfed farming areas generally emphasis is on

(a) Moisture conservation practices
(b) Drainage
(c) Disposal of excess water
(d) All of these

35. Higher yield of rice could be ensured by adequate supply of ______ fertilizer at the tollering stages

(a) Nitrogen (b) Potash
(c) Phosphorus (d) None of these

36. The areas where drought has occurred in_______ of years during the period are considered as drought areas

(a) 60 % (b) 40 %

(c) 30 % (d) 10%

37. What sort of aberration in rainfall affects crop growth most

(a) Lesser rainfall (b) Higher rainfall

(c) Uncertain distribution (d) None of these

38 Which of these influence most the water requirement of a crop:

(a) Weather (b) Soil condition

(c) Water table (d) All of these

39. What is dryland ?

(a) When rainfall received is less than 1000 mm

(b) When rainfall received is less than 500 mm

(c) When rainfall received is less than 750 mm

(d) Hilly terrain with low rainfall less than 500 mm

40. In undulating terrain one should go for________

(a) Normal crop raising (b) Strip cropping

(c) Pastoral cropping (d) Tree cropping

41. Atrazine used as an antitranspirant is order to

(a) Reduces the growth of the crop

(b) Does not reflect light from plant leaf surface

(c) Affects the closure and opening of stomata

(d) None of these

42. Phenylmercuric acetate (PMA) is a chemical used in agricultural crops in order to

(a) Increase CO_2 uptake (b) Increase transpiration

(c) Reduce transpiration (d) None of these

43. Water requirement (WR) includes the losses due to

(a) Evapotranspiration and application of water

(b) Consumptive use, water required for special operations and other economically unavoidable losses of water.

(c) ET and water required for special operations.

(d) None of these

44. Dust mulch is very effective in

(a) Coarse textured soil (b) Heavy textured soil

(c) Sandy soil (d) All of these

45. The characteristic of dryland crop is

(a) Shorter duration
(b) Faster growth
(c) Strong root system
(d) All of these

46. Which of the following is not a characteristic of dryland crop ?

(a) Potential for high yield
(b) Fertilizer responsive
(c) Storage root system
(d) High water requirement

47. Which is the most popular dryland crop ?

(a) Sorghum
(b) Rice
(c) Wheat
(d) Mustard

48. Soil mulch reduces

(a) Deep cracking
(b) Reduces evaporation
(c) Improves water storage
(d) All of these

49. What is Hedgerow intercropping ?

(a) Cultivation of food crops in alleys formed by hedge row of trees /shrubs in arable lands
(b) Cultivation of food crops in alleys formed by hedge row of trees/ shrubs in waste lands
(c) Cultivation of food crops in alleys formed by hedge row of trees/ shrubs in forest lands
(d) None of these

50. Alley system is a

(a) Forage-alley cropping
(b) Forage-cum-mulch system
(c) Forage-cum-pole system
(d) All of these

51. The plastic mulch is helpful in

(a) Reduce soil moisture
(b) Increase the evaporation
(c) Increase soil temperature
(d) Better crop

52. A year in which rainfall of a particular place is short by more than________ is known as drought year.

(a) Twice the mean deviation
(b) Twice the standard deviation
(c) Twice the median
(d) None of these

53. Abnormally rapid erosion in an environment disturbed by animal life, chiefly man is called

(a) Erodibility (b) Erosion resistant
(c) Erosion accelerated (d) None of these

54. A crop, because of its dense foliage, root system etc., provides effective protection against soil loss by erosion is called

(a) Erodibility (b) Erosion resistant crop
(c) Erosion accelerated (d) Erosion permitting crops

55. A crop, which are increased the soil losses by erosion is known as

(a) Erosion permitting crops (b) Erosion resistant crops
(c) Erosion accelerated (d) None of these

56. Water erosion producing very small and numerous channels is known as

(a) Gully erosion (b) Sheet erosion
(c) Rill erosion (d) None of these

57. Supplemental irrigation can be given to save the crop in

(a) Dry farming (b) Dry season
(c) Drought year (d) Organic farming

58. No supplemental irrigation is required to save the crop in

(a) Rainfed farming (b) Dry season
(c) Drought year (d) None of these

59. Drier part of Rajasthan, Gujarat, Maharashtra etc. comes under

(a) Dry farming (b) Organic farming
(c) Drought year (d) Rainfed farming

60. Humid and sub humid part of north eastern states comes under

(a) Organic farming (b) Dry farming
(c) Drought year (d) Rainfed farming

61. Conservation of rain water under unirrigated conditions, by collecting runoff of precipitation in order to supplement soil moisture in an adjacent area is known as

(a) Rain water harvesting (b) Soil conservation
(c) Runoff harvesting (d) None of these

62. A combination of all management and land use methods which safe guard the soil against depletion by natural or by man induced factors is known as

(a) Water harvesting (b) Soil conservation
(c) Runoff harvesting (d) Watershed

63. **The area above a given point on a stream that contributes water to the flow at that point is called:**
(a) Watershed (b) Soil conservation
(c) Runoff harvesting (d) None of these

64. **An implement used for puddling the paddy field before transplanting is:**
(a) Rotavotor (b) Harrow
(c) Ridge & furrow makers (d) Puddler

65. **Which of the following is/are water absorbing chemicals ?**
(a) Grow soak (b) Agrolite-400
(c) Jalshakti (d) All of these

66. **It is the force with which the water is held by soil against gravity is known as**
(a) Soil moisture deficit (b) Soil moisture stress
(c) Soil moisture tension (d) None of these

67. **Incorporation of vegetative mulch in a band in the soil for the purpose of harvesting rain water and conserving soil and water is known as**
(a) Tillage action (b) Tillage depth
(c) Tillability (d) Vertical mulching

68. **A form of soil erosion resulting from soil splash under the impact of falling rain drops is known as:**
(a) Splash erosion (b) Gully erosion
(c) Rill erosion (d) None of these

69. **Erosion of a fairly uniform layer of material from the land surface, often scarcely perceptible especially when caused by wind is**
(a) Gully erosion (b) Sheet erosion
(c) Rill erosion (d) None of these

70. **Removal of soil by running water, with formation of channels, that cannot be smothed out completely by normal cultivation is known as**
(a) Gully erosion (b) Sheet erosion
(c) Rill erosion (d) None of these

71. **Climate under crop canopy is known as**
(a) Eco climate (b) Crop climate
(c) Both of these (d) None of these

72. **The amount of water vapour in the air is called:**
(a) Humidity (b) Frost
(c) Mist (d) Ice

73. A chemical that selectivity interferes with normal hormonal production of growth but without appreciable toxic effect is called:

(a) Growth regulator (b) Growth retardant
(c) Growth stimulant (d) None of these

74. The zone delineated by Thornthwaite's moisture indices ranges from -20 to -40 is known as

(a) Semi-arid (b) Arid
(c) Humia (d) None of these

75. The zone delineated by Thornthwaites's moisture indices ranges from -60 to -80 is known as

(a) Semi-arid (b) Arid
(c) Humid (d) None of these

76. The zone delineated by Thornthwaite's moisture indices varying from less than -80 is:

(a) Semi-arid (b) Arid
(c) Severe arid (d) Extreme humid

77. The zone delineated by Thornthwaite's moisture indices varying from 20 to 50 is:

(a) Semi-arid (b) Arid
(c) Severe arid (d) Humid

78. The zone delineated by Thornthwaite's moisture indices ranges from 50 to 100 is:

(a) Semi- arid (b) Arid
(c) Severe humid (d) Humid

79. Measuring depth of a water from precipitation is

(a) Raingauge (b) Sunshine recorder
(c) Anemometer (d) Barometer

80. The period of the year in which monthly rainfall exceeds 10 per cent of the monthly evapotranspiration is known as

(a) Rainy season (b) Dry season
(c) Wet season (d) None of these

81. A long, low ridge of earth with gently sloping sides and a shallow channel along the upper side, to control erosion by diverting surface runoff across the slope instead of permitting it to flow uninterrupted down to slope is called:

(a) Ridge terrace (b) Ridge planting
(c) Rotary tillage (d) None of these

82. **Planting on ridges is known as**

(a) Ridge terrace (b) Ridge planting
(c) Rotary tillage (d) None of these

83. **A kind of drought which occurs during distinct annual periods of dry weather is known as**

(a) Seasonal drought (b) Recurrent drought
(c) Both (a) and (b) (d) None of these

84. **A land or climate where the rainfall is greater many crops. Irrigation or dryland farming methods are commonly required for crop production is**

(a) Arid (b) Semi arid
(c) Subhumid (d) None of these

85. **The value of albedo in the dryland soils ranges from**

(a) 0.10-0.25 (b) 0.80 - 0.90
(c) 0.50 - 0.75 (d) 0.01 - 0.20

86. **Generally drought is assessed using**

(a) Aridity (b) Dryness
(c) Aridity anomaly index (d) Deficient rainfall

87. **Materials which reduces water loss from leaf surface by reducing the size or number of stomatal openings are called**

(a) Growth retardants (b) Growth promoters
(c) Anti-transpirants (d) All of these

88. **Which is a stomata closure type of anti-transpirants ?**

(a) PMA (b) Kaolin
(c) Mobileaf (d) None of these

89. **Which is a film forming type of anti-transpirants ?**

(a) PMA (b) Kaolin
(c) Mobileaf (d) None of these

90. **Which is a leaf reflectance type of anti-transpirants ?**

(a) PMA (b) Kaolin
(c) Mobileaf (d) None of these

91. **The departure of aridity index from normal by ____ represents moderate drought**

(a) <0.5 (b) 0.5-1.0
(c) 1.0-1.5 (d) >3.0

92. The departure of aridity index from normal by _____ represents disastrous drought

(a) <0.5 (b) 0.5-1.0
(c) 1.0-2.0 (d) >2

93. The departure of aridity index from normal by_____ represents large drought

(a) <0.5 (b) 0.5-1.0
(c) 1.0-2.5 (d) >3

94. The departure of aridity index from normal by_____ represents severe drought

(a) <0.5 (b) 0.5-1.0
(c) 1.0-2.0 (d) 3.0-5.0

95. The drought zones areas with _______ probability of rainfall departure from normal

(a) 25% (b) 45%
(c) 30% (d) 60%

96. The dry farming research station was started in the year 1934 at

(a) Hagari (Karnataka) (b) Raichur (Karnataka)
(c) Manjri (Maharashtra) (d) Both (a) & (b)

97. The dry farming research station was started in the year 1933

(a) Bijapur (Karnataka) (b) Hagari (Karnataka)
(c) Manjri (Maharashtra) (d) Rohtak (Haryana)

98. The first scientific approach to tackle the problem of dry farming areas was initiated by

(a) B.P. Pal (b) M.S. Swaminathan
(c) V.A. Tamhane (d) R.S. Singh

99. Which of the following is a permanent climatic feature and is the culmination of long term process ?

(a) Aridity (b) Drought
(c) Dryland (d) All of these

100. Which is a condition when the actual seasonal rainfall is deficient by more than twice the mean deviation ?

(a) Aridity (b) Drought
(c) Dryland (d) All of these

101. What is Dryfarming ?

(a) When rainfall received is less than 1000 mm
(b) When rainfall received is less than 400 mm
(c) When rainfall received is less than 750 mm
(d) None of these

102. What is Rainfed farming area?

(a) When rainfall received is more than 1150 mm
(b) When rainfall received is less than 500 mm
(c) When rainfall received is less than 800 mm
(d) None of these

103. Dryfarming areas are characterized by a growing season of

(a) <75 days (b) 75-120 days
(c) >120 days (d) None of these

104. Dryland farming areas are characterized by a growing season of

(a) <75 days (b) 75-120 days
(c) >120 days (d) All of these

105. By delayed planting of rice in the second season the yield decreases most due to ______

(a) Poor tillering
(b) Higher temperature at flowering
(c) Heavy insect infestation
(d) None of these

106. The First Dry Farming Research Station started in the year 1923 at

(a) Bijapur (b) Nagpur
(c) Jaipur (d) Manjri

107. Dry Farming Research Station started in the year 1935 at

(a) Bijapur (b) Sholapur
(c) Manjri (d) Rohtak

108. The drought found in climates with well defined rainy and dry seasons is known as

(a) Invisible (b) Permanent
(c) Seasonal (d) None of these

109. The drought characterized by abnormal failure of rainfall is known as

(a) Invisible (b) Contingent
(c) Seasonal (d) None of these

110. When rainfall is inadequate to meet the ET looses and result in borderline deficiency of water in soil resulting in less than quantum yield is known as

(a) Invisible (b) Seasonal
(c) Permanent (d) None of these

111. When annual precipitation is less than the normal over an area for prolonged period is known as

(a) Meteorological drought (b) Atmospheric drought
(c) Hydrological drought (d) Socioeconomic drought

112. Drought due to low air humidity and accompanied by hot dry wind is

(a) Socio-economic drought (b) Atmospheric drought
(c) Hydrological drought (d) Agricultural drought

113. The chronic drought zones areas with______ probability of rainfall departure from normal

(a) 25% (b) 30%
(c) 40% (d) 60%

114. Which is not a moisture conservation practice in drylands ?

(a) Tillage (b) Fallowing
(c) Mulching (d) None of these

115. Vertical mulching has been adopted in which types of soils ?

(a) Heavy soils (b) Light soils
(c) Loamy soils (d) Sandy soils

116. Trenches of 30x60 cm are dug at 5-6 m interval and filled with stalk to keep the cracks open for better water intake in

(a) Vertical mulching (b) Dust mulch
(c) Stubble mulch (d) None of these

117. Which of the following are stomata closure type of anti-transpirants ?

(a) Hexadecanol (b) Silicon
(c) Mobileaf (d) All of these

118. CCC is a which type of anti-transpirants ?

(a) Film forming (b) Stomatal closure
(c) Growth retardant (d) All of these

119. The process of runoff collection during peak periods of rainfall in storage tanks and ponds is known as

(a) Water harvesting (b) Runoff collection
(c) Water storage (d) All of these

120. In agri-horticulture system for arable land which is the tree component ?

(a) Fruit (b) Timber
(c) Hedgerow (d) None of these

121. Leaf surface coating with white material increases

(a) Albedo (b) Stomatal closure
(c) Growth (d) None of these

122. Which type of anti-transpirants reduce transpiration loss by reducing leaf temperature and vapour pressure gradient ?

(a) Film forming (b) Stomatal closure
(c) Leaf reflectance (d) All of these

123. Which type of anti-transpirants produce external physical barrier to reduce vapour loss from leaf surface ?

(a) Film forming (b) Stomatal closure
(c) Leaf reflectance (d) All of these

124. Which is at low concentrations also serves as stomata ?

(a) 2,4-D (b) MCPA
(c) Atrazine (d) Kaolin

125. TIMFIB represents

(a) Timber cum fiber system (b) Timber cum food system
(c) Timber cum fodder system (d) All of these

126. Moisture stress in soils and plants during crop growth due to imbalance between soil moisture and evapotranspiration of a crops is __________

(a) Agricultural drought (b) Hydrological drought
(c) Meteorological drought (d) None of these

127. Antitranspirant belongs to stomatal closing group is

(a) Cycocel (b) PMA
(c) 2,4-D (d) Silicon

128. Which one of the following chemical is used as anti transpiration in dryland crops ?

(a) Paraqual (b) MCPA
(c) Cycocel (d) None of these

129. Which is the most appropriate classification scheme which relates climate to vegetation ?

(a) Penman classification (b) Thomthwite classification
(c) Koppens classification (d) None of these

130. Drought has been classified into______

(a) 2 types (b) 3 types
(c) 4 types (d) All of these

131. Which of the following is not a kind of drought ?

(a) Invisible (b) Climatic
(c) Seasonal (d) Permanent

132. Desert climate is the characteristic of _____

(a) Invisible drought (b) Climatic drought
(c) Seasonal drought (d) Permanent drought

133. Sparse vegetation adopted to drought and crop raising only through irrigation during entire period is the characteristic of ________

(a) Invisible drought (b) Climatic drought
(c) Seasonal drought (d) Permanent drought

134. Meteorological drought for prolonged period with depletion of surface water and consequent drying of reservoirs is

(a) Meteorological drought (b) Socioeconomic drought
(c) Hydrological drought (d) Agricultural drought

135. Drought due to imbalance between available soil moisture and ET of a crop is

(a) Socioeconomic drought (b) Atmospheric drought
(c) Hydrological drought (d) Agricultural drought

136. Which of the following is not a plant adaptation to drought

(a) Escaping drought (b) Drought resistance
(c) Drought tolerance (d) Drought survival

137. The concept of aridity index was given by

(a) Henry Gillbert (b) C.W. Thronthwaite
(c) Phillip Tow (d) V. A. Tamhane

138. Horti/ silvi-pastural system is suitable for

(a) Atable lands (b) Non arable lands
(c) Hilly lands (d) All of these

139. Horti/ silvi-pastural system is suitable for

(a) Class 4 and above (b) Class 4 and below land
(c) Hilly land (d) All of these

140. Growing of pasture for grazing and conservation is known as

(a) Alternate husbandary (b) Mixed farming
(c) Both (a) and (b) (d) Ley farming

141. Area covered with pasture which are reseeded at regular intervals is termed as

(a) Alternate husbandary (b) Mixed farming
(c) Organic farming (d) Ley farming

142. What is ley farming ?

(a) Cultivation of pasture in arable lands
(b) Cultivation of pasture in non arable lands
(c) Cultivation of food pasture in forest lands
(d) All of these

143. Usually crop face moisture stress during their life cycle in case of

(a) Dry farming (b) Dry season
(c) Drought year (d) Organic farming

144. Rainfall received less than 750 mm during their life cycle in case of

(a) Dry farming (b) Dry season
(c) Drought year (d) None of these

145. Rainfall received more than 750 mm during their life cycle in case of

(a) Dry farming (b) Dry season
(c) Drought year (d) Organic farming

146. Over flow weir is also known as:

(a) Check dams (b) Field bunds
(c) Terraces (d) None of these

147. Soil erosion can be caused by:

(a) High plant density (b) Over grazing
(c) Terracing (d) All of these

148. Soil temperature can be regulated by use of straw mulch both in summer and winter season because:

(a) It reduces evaporation
(b) It lowers the soil temperature
(c) It keeps the temperature of soil high
(d) It provides insulating effect

149. Gully erosion of soil is caused by:

(a) Stream formed by rain water (b) Wind
(c) Terrace farming (d) All of these

150. Which one of the following can lead to soil erosion ?

(a) Contour ploughing (b) Terrace farming
(c) Deforestation (d) Organic farming

151. Which one is an example of drought resistant crop ?

(a) Sorghum (b) Rice
(c) Barley (d) Potato

152. Which chemical regulates opening and closing of stomata?

(a) 2, 4-4 (b) PMA
(c) IAA (d) Kaoline

153. The Central Research Institute for Dryland Agriculture (CRIDA) is located at

(a) Nagpur (b) Cuttack
(c) Hyderabad (d) New Delhi

154. Physiological drought is mainly caused by_______

(a) Excess water (b) Plant pathogen
(c) Salinity and alkalinity (d) All of these

155. Maximum wind soil erosion occurs in _______

(a) Maharashtra (b) Karnataka
(c) Uttar Pradesh (d) Rajasthan

156. Which one of the following is the first knane of water erosion ?

(a) Splash erosion (b) Rill erosion
(c) Gully erosion (d) Sheet erosion

157. Bund construction across the slopes are known as:

(a) Gully plugs (b) Torraces
(c) Contour bunds (d) None of these

158. Series of flat plateform constructed across a slopping land which give rise to the formation of a step or bench like feature is known as:

(a) Gully plugs (b) Terraces
(c) Contour bunds (d) None of these

159. Small earthen barriers constructed in field is known as:

(a) Gully plugs (b) Contour bunds
(c) Field bunds (d) None of these

160. Construction of field bunds along the contour line is known as:

(a) Contour bunds (b) Terraces
(c) Field bunds (d) None of these

161. Geological erosion is also known as:

(a) Water erosion (b) Natural erosion
(c) Soil erosion (d) Wind erosion

162. In which one of the following states is the terrace cultivation practice ?

(a) Bihar (b) Haryana
(c) Rajasthan (d) Uttarakhand

Fill in the blanks

1. Permits a crop to grow for a longer time in a given moisture stress conditions is called__________
2. Climate under crop canopy is called________
3. The year in which rainfall of particular place is short by more than twice the standard deviation is called_______

4. A period of at least 15 consecutive days, during which not a single wetting rain has fallen is called______
5. A kind of drought which occurs during distinct annual periods of dry weather is called ________
6. Supplemental irrigation can be given to save the crop is __________
7. Drier part of Rajasthan, Gujarat, Maharashtra etc. come under_______
8. Planting on ridges is called_______
9. A dry soil layer at the surface may be created *in-situ* by loosening the surface soil and hastening the drying of the upper soil is known as ______
10. The International Centre for Agricultural Research in Dry Areas (ICARDA) was established in the year ________
11. Leaf and canopy temperature can be measured by_____
12. Water erosion producing very small and numerous channels is called ______
13. The amount of water vapour in the air is called____
14. Geological erosion is also known as _____
15. Contour bunding and terracing are the measures of _____

True/ False

1. The lack of resistance of a soil to soil erosion is called erodibility. **(T)**
2. A crop, which are increased the soil losses by erosion is called erosion resistant crops. **(F)**
3. Splash erosion is the first of water erosion. **(T)**
4. Black plastic is more effective in mulching. **(T)**
5. Geological erosion is also known as water erosion. **(F)**
6. Mulching is a process that helps mainly in weed control. **(F)**
7. Water absorbing chemicals are grow soak. **(T)**
8. Maximum wind soil erosion occurs in West Bengal. **(F)**
9. Sorghum is drought resistant crop. **(T)**
10. Centrol Arid Zone Research Institute is situated at Jodhpur. **(T)**

Answers Fill in the Blanks

1. Drought prone
2. Ecoclimate
3. Drought year
4. Dry farming
5. Seasonal drought
6. Dry Farming
7. Dry Farming
8. Ridge planting
9. Soil mulch
10. 1977
11. Infra red thermometer
12. Rill erosion
13. Humidity
14. Natural erosion
15. Mechanical measures

ANSWERS

1.(a)	2.(d)	3.(d)	4.(c)	5.(d)	6.(d)	7.(a)
8.(a)	9.(b)	10.(c)	11.(a)	12.(b)	13.(a)	14.(a)
15.(a)	16.(b)	17.(b)	18.(b)	19.(b)	20.(b)	21.(c)
22.(a)	23.(c)	24.(a)	25.(b)	26.(c)	27.(a)	28.(b)
29.(c)	30.(d)	31.(c)	32.(a)	33.(d)	34.(c)	35.(a)
36.(d)	37.(c)	38.(a)	39.(a)	40.(b)	41.(a)	42.(c)
43.(c)	44.(b)	45.(d)	46.(d)	47.(a)	48.(d)	49.(a)
50.(d)	51.(c)	52.(a)	53.(c)	54.(b)	55.(a)	56.(c)
57.(a)	58.(d)	59.(a)	60.(d)	61.(a)	62.(b)	63.(a)
64.(d)	65.(d)	66.(c)	67.(d)	68.(a)	69.(b)	70.(a)
71.(a)	72.(a)	73.(b)	74.(a)	75.(b)	76.(c)	77.(d)
78.(c)	79.(a)	80.(a)	81.(a)	82.(b)	83.(a)	84.(b)
85.(c)	86.(a)	87.(d)	88.(a)	89.(c)	90.(b)	91.(a)
92.(d)	93.(b)	94.(c)	95.(a)	96.(d)	97.(a)	98.(c)
99.(a)	100.(b)	101.(c)	102.(c)	103.(a)	104.(b)	105.(b)
106.(d)	107.(d)	108.(c)	109.(b)	110.(a)	111.(a)	112.(b)
113.(c)	114.(d)	115.(a)	116.(a)	117.(d)	118.(c)	119.(a)
120.(a)	121.(a)	122.(c)	123.(a)	124.(c)	125.(a)	126.(a)
127.(b)	128.(c)	129.(b)	130.(c)	131.(b)	132.(d)	133.(d)
134.(c)	135.(d)	136.(d)	137.(b)	138.(b)	139.(a)	140.(c)
141.(d)	142.(b)	143.(a)	144.(a)	145.(a)	146.(a)	147.(b)
148.(d)	149.(a)	150.(c)	151.(a)	152.(b)	153.(c)	154.(c)
155.(d)	156.(c)	157.(a)	158.(b)	159.(c)	160.(a)	161.(d)
162.(d)						

11

Cereal Crop Production

1. **The centre of origin of wheat is**
   (a) S-W Asia (b) India
   (c) S. Africa (d) Mexico
2. **In India the productivity of wheat is highest in the state**
   (a) Punjab (b) Haryana
   (c) W. Bengal (d) U.P.
3. **The inflorescence in wheat is**
   (a) Siliqua (b) Panicle
   (c) Spike (d) Raceme
4. **Area under wheat cultivation in India is approximately**
   (a) 49 m ha (b) 45 m ha
   (c) 40 m ha (d) 29 m ha
5. **Optimum temperatures towards maturity of wheat lies between**
   (a) 21-30°C (b) 25-30°C
   (c) 30-35°C (d) 18-22°C
6. **Which is a suitable variety of wheat under rainfed conditions ?**
   (a) C 306 (b) K 65
   (c) Narbada 4 (d) All of these
7. **Which is a durum variety of wheat ?**
   (a) PBW 34 (b) Malvika
   (c) Jairaj (d) All of these
8. **Which is a single gene dwarf variety of wheat ?**
   (a) HD 2177 (b) HD 2189
   (c) Narbada 4 (d) PBW-34

9. Which is a double gene dwarf variety of wheat ?

(a) HD 2177 (b) HD 2189
(c) Narbada 4 (d) Jairaj

10. The recommended seed rate for wheat is ________

(a) 80-100 kg/ha (b) 15-20 kg/ha
(c) 50-70 kg/ha (d) 30-40 kg/ha

11. Higher yield of wheat could be achieved by

(a) Use of higher dose of phosphate (b) Adequate amount of N
(c) No nitrogen application (d) All of these

12. The second most cultivated crop of the world is

(a) Wheat (b) Rice
(c) Maize (d) Gram

13. The recommended dose of N, P_2O_5 and K_2O kg/ha for dwarf wheat are

(a) 100-60-40 (b) 120-40-40
(c) 30-20-30 (d) 40-30-40

14. The N fertilizer use efficiency in wheat can be increased by using

(a) S-coated urea (b) Urea super granules
(c) Blue Green Algae (d) Both (a) and (b)

15. Triticale is a cross between

(a) Wheat x rye (b) Oat x barley
(c) Wheat x barley (d) Wheat x oat

16. *Phalaris minor* belongs to the family

(a) Cyperacene (b) Gramineae
(c) Leguminaceae (d) Euphorbiaceae

17. The nitrogen losses in wheat can be reduced by

(a) Basal application (b) Top dressing
(c) Split application (d) Both (b) and (c)

18. First irrigation to the wheat crop should be given at _____ stage

(a) 21 days (b) 40 days
(c) 60 days (d) 70 days

19. *Phalaris minor* weed in wheat can be controlled by using

(a) 2, 4-D (b) Isoproturon
(c) Atrazine (d) None of these

20. Test weight of *Phalaris minor* is

(a) 2 g (b) 8 g
(c) 22 g (d) 20 g

21. Cultivation of macaroni wheat is restricted to

(a) Western India
(b) Central & Southern India
(c) North-eastern India
(d) None of these

22. Wheat species known as *Triticum sphaerococcum* has been put out of cultivation because of

(a) Low productivity and high susceptibility of disease
(b) Bad taste and bad cooking quality
(c) Quick spoilage in the shortage
(d) All these reasons

23. *Triticum aestivum,* the common bread wheat which is the most important wheat species in India, occupies the following proportion of India's total wheat area

(a) 24%
(b) 85%
(c) 75%
(d) 69%

24. Which of the following species of wheat are cultivated all over India from the sea level upto high snow cover red regions of Himalaya is

(a) *Triticum dicoccum*
(b) *Triticum durum*
(c) *Triticum aestivum*
(d) *Triticum sphaerococum*

25. *Triticum durum,* the macaroni wheat, occupies the following proportion of India's cultivated area under wheat

(a) 14%
(b) 4%
(c) 75%
(d) 30%

26. Wheat species known as *Triticum dicoccum* is cultivated on a very restricted scale in

(a) Gujarat
(b) Maharashtra
(c) Andhra and Tamil Nadu
(d) All of these

27. Optimum temperatures for growth and development of wheat ranges between

(a) 21-37°C
(b) 25-30°C
(c) 10-15°C
(d) 15-20°C

28. Optimum temperatures for spouting of wheat is

(a) 15°C
(b) 20-25°C
(c) 4-5°C
(d) 15-20°C

29. Temperatures having detrimental effect on growth of wheat is

(a) 15-25°C
(b) 25-30°C
(c) 30-35°C
(d) 20-25°C

30. The shoot of wheat is known as

(a) Stem (b) Trunk
(c) Culm (d) None of these

31. Shoot of wheat is

(a) Hollow (b) Solid
(c) Hollow except at nodes (d) Soild except at nodes

32. The chromosome numbers in *Triticum aestivum* is

(a) 21 (b) 9
(c) 17 (d) 41

33. The chromosome numbers in *Triticum monococcum* is

(a) 21 (b) 7
(c) 14 (d) None of these

34. The chromosome numbers in *Triticum durum* is

(a) 21 (b) 7
(c) 14 (d) None of these

35. Common bread wheat is

(a) Diploid (b) Tetraploid
(c) Hexaploid (d) All of these

36. Emmer wheat is

(a) *Triticum aestivum* (b) *Triticum durum*
(c) *Triticum vulgare* (d) *Triticum dicoccum*

37. Macaroni wheat is

(a) *Triticum aestivum* (b) *Triticum durum*
(c) *Triticum vulgare* (d) *Triticum dicoccum*

38. Which of the following is most common weed in wheat ?

(a) *Phalaris minor* (b) *Avena fatua*
(c) *Cyperus rotundus* (d) *Argemone maxicana*

39. Which one of the following is the characteristic of stripe rust of wheat ?

(a) Brown, elongated ruptured pustules on stem and leaf-sheath
(b) Yellow, small, spherical to oval ruptured uredo-pustules in rows of leaf
(c) Brown, spherical to oval, ruptured uredo-pustules on leaf
(d) None of these

40. What is optimum moisture content for storing wheat ?

(a) 10% (b) 12%
(c) 14% (d) 16%

41. Which of the following is the ideal temperature range for tilllering stage in wheat ?

(a) 10-15°C (b) 16-20°C
(c) 20-25°C (d) 23-30°C

42. Species of wheat not cultivated at present in India is

(a) *Triticum aestivum* (b) *Triticum dicoccum*
(c) *Triticum durum* (d) *Triticum sphaerococum*

43. Leaf of wheat is

(a) Sessile (b) Petiolate
(c) Both (a) and (b) (d) None of these

44. In wheat ______ is very prominent

(a) Auricle (b) Ligule
(c) Lemma (d) All of these

45. Panicle of wheat is borne on

(a) Uppermost node (b) Lowermost node
(c) Middle nodes (d) None of these

46. Hull of wheat comprised of

(a) Palea (b) Lemma
(c) Both (b) and (c) (d) None of these

47. The directorate of Wheat Research Institute is situated at ______

(a) Karnal (b) Coimbatore
(c) Lucknow (d) New Delhi

48. The optimum plant population for wheat is around

(a) 1-2 lacs (b) 2-3 lacs
(c) 3-4 lacs (d) 4-5 lacs

49. Wherever one irrigation is available, the wheat crop should irrigated at ______ stage

(a) CRI (b) Tillering
(c) Flowering (d) Milking

50. Which one is the most critical stage of irrigation in wheat ?

(a) Tillering (b) CRI
(c) Milking (d) Flowering

51. The normal seed rate of wheat is

(a) 40 kg/ha (b) 60 kg/ha
(c) 100 kg/ha (d) 120 kg/ha

52. **Broad leaved weeds in wheat can be controlled by applying**
 (a) Isoproturon (b) 2,4-D
 (c) Metsulfurom methyl (d) All of these

53. **Which is the most cultivated crop of the world ?**
 (a) Rice (b) Wheat
 (c) Maize (d) Barley

54. **Which is the second most cultivated crop in India?**
 (a) Rice (b) Wheat
 (c) Maize (d) Barley

55. **Which one of the following statements refers to the concept of critical stage of irrigation :**
 (a) When irrigation is not given reduction of yield is maximum
 (b) When water absorption rate is maximum
 (c) When water loss by the plants is maximum
 (d) All of the above

56. **Wheat variety _______ has given consistently superior performance in Barani (rainfed areas) tracts of the country.**
 (a) N. P. 875 (b) Sonalika
 (c) C- 306 (d) All of these

57. **Wheat varieties cultivated in India belongs to species**
 (a) Aestivum (b) Durum
 (c) Dicocum (d) None of these

58. **High yielding dwarf varieties of wheat were developed by**
 (a) Dr. Norman E. Borlaug (b) Dr. B. P. Pal
 (c) Dr. M.S. Swaminathan (d) None of these

59. **The *triticum aestivum* is ________ wheat .**
 (a) Hexaploid (b) Diploid
 (c) Tetraploid (d) None of these

60. **The optimum temperatures range for sowing of wheat is**
 (a) 15-20°C (b) 12-15°C
 (c) 20-25°C (d) None of these

61. **Optimum temperatures throughout the growth of rice lies between**
 (a) 21-37°C (b) 25-30°C
 (c) 20-25°C (d) 15-24°C

62. **Optimum temperatures for blooming of rice lies between**
 (a) 21.5-37°C (b) 25-30.5°C
 (c) 26.5-29°C (d) 15-19°C

63. Minimum temperatures for sprouting of rice is

(a) 15°C (b) 22-23°C
(c) 10°C (d) 17°C

64. Minimum temperatures for flowering of rice is

(a) 15-17°C (b) 25-30°C
(c) 10-15°C (d) 18-20°C

65. The shoot of rice is known as

(a) Stem (b) Trunk
(c) Culm (d) None of these

66. The varieties which belong to species *Oryza glaberrima* are found in

(a) China (b) India
(c) America (d) Africa

67. Rice grown in India belongs to

(a) Indica (b) Japonica
(c) Asiatica (d) None of these

68. The rice grown in Indonesia belongs to

(a) *Oryza glaberrima* (b) *Oryza sativa indica*
(c) *Oryza sativa japonica* (d) *Oryza sativa javanica*

69. Non traditional areas of rice cultivation include

(a) Punjab and Haryana (b) Uttar Pradesh and Bihar
(c) West Bengal and Odisha (d) Tamil Nadu and Haryana

70. The rice inflorescence is

(a) Panicle (b) Spikelets
(c) Raceme (d) Siliqua

71. Dead heart and white head damage to rice is caused by

(a) Gall midge (b) Leaf roller
(c) Gundhi bug (d) Stem borer

72. Silveryshoot or onion leaf symptom of rice tillers

(a) Army worm (b) Gall midge
(c) Leaf roller (d) None of these

73. Grassy stunt virus disease of rice is transmitted by

(a) Green leaf hopper (b) Brown plant hopper
(c) White fly (d) Gall midge

74. Gundhi bug insect attack on rice during

(a) Germination (b) Tillering stage
(c) Milking stage (d) Harvesting stage

75. **Test weight of basmati rice is**

(a) 17 g (b) 21 g
(c) 28 g (d) 32 g

76. ***Oryza sativa* is a diploid species having chromosome**

(a) 11 (b) 24
(c) 46 (d) None of these

77. **In Orissa summer or spring season rice is known as**

(a) Boro (b) Arnall
(c) Dalua (d) Kodo

78. **Dalua crop is sown in**

(a) Nov-Dec (b) June-July
(c) May-June (d) February-March

79. **Boro crop is sown in**

(a) Nov-Dec (b) July-August
(c) May-June (d) March-April

80. **Aman crop is sown in**

(a) Dec-January (b) June - July
(c) May-June (d) March - April

81. **Aus crop is sown in**

(a) Dec- January (b) July- August
(c) May-June (d) March - April

82. **Number of species included under genus *oryza***

(a) 23 (b) 21
(c) 17 (d) 24

83. **Rice varieties cultivated in India belongs to genus**

(a) *Asiatica* (b) *Japonica*
(c) *Javanica* (d) None of these

84. **Optimum temperatures towards maturity of rice lies between**

(a) 21-37°C (b) 25-30°C
(c) 20-25°C (d) 15-20°C

85. **Which is a long duration variety of rice ?**

(a) Govind (b) Bala
(c) Garima (d) None of these

86. **Which is an early maturing semi dwarf variety of rice ?**

(a) Govind (b) Bala
(c) Garima (d) None of these

87. Most serious pest of wheat is

(a) Termite (b) Cockroach
(c) Moth (d) Leaf hoppers

88. The panicle initiation stage in rice plants comes

(a) Immediate after transplanting (b) After maximum tillering stage
(c) After boot leaf stage (d) At maturity stage

89. Which gas is released from paddy fields

(a) CH_4 (b) H_2S
(c) NH_3 (d) Chlorine

90. Varieties of rice cultivated in India belong to

(a) *Oryza sativa* (b) *Oryza glaberrima*
(c) *Oryza perennis* (d) None of these

91. Seed of rice is known as

(a) Siliqua (b) Caryopsis
(c) Droop (d) None of these

92. When rice seedlings are raised in puddle beds, it is known as

(a) Wet nursery (b) Field nursery
(c) Dapog nursery (d) None of these

93. When rice nursery is grown in raised beds of well prepared dry soils (without stagnant water) , it is called

(a) Raised nursery (b) Dry nursery
(c) Dapog nursery (d) Puddle nursery

94. *Rabi* rice seedlings are transplanted at the age of

(a) 10-16 days (b) 30-40 days
(c) 45-50 days (d) None of these

95. Which of the following rice varieties are not drought tolerant

(a) N-22, Bala, Brown Gora, White Gora, Black Gora, Rasi
(b) Lal Nakanda, CR 143-2-2
(c) CO 14, Vani, K 333
(d) None of these

96. Blast resistant varieties of rice are

(a) CO 14, K 332 (b) K 333, Jaya
(c) Rasi, Vani (d) All of these

97. Rice varieties resistant to bacterial blight are

(a) IET-4141 (b) BJ-1
(c) TKM 6 (d) All of these

98. Which of the following is stem borer tolerant varieties of rice ?

(a) TKM 6, IET 2815
(b) W 1263, IET 2845
(c) IET 2812
(d) All of these

99. Which of the following rice varieties resistant to white backed plant hopper ?

(a) T- 1417, Lal Basmati
(b) Eswaramangalam
(c) Kodiyan
(d) All of these

100. Which one of the following rice varieties resistant to brown plant hopper ?

(a) PTB 21, PTB 33, IET 5118
(b) IET 5119, T-1425, N - 1432
(c) Ennapatta, T- 1421, Sonamukhi, Pandi, ARC 6650
(d) All of these

101. Rice varieties which are resistant to tungru virus is

(a) Kataribhog, PTB 2
(b) TKM 6
(c) IR 362
(d) None of these

102. Rice varieties which are resistant to grassy-stunt virus is

(a) O. nivari, IR 362
(b) Kataribhog
(c) TKM 6
(d) None of these

103. Which of the following rice varieties resistant to green leaf hopper ?

(a) Vani, PTB 2, PTB 7
(b) RP 5-12, Pankhari 203
(c) PTB 18
(d) All of these

104. Which one of the following rice varieties resistant to gall midge ?

(a) Eswarakora, Surek
(b) CR 94 , MRI 523
(c) PTB 18, PTB 21, P
(d) All of these

105. Photoperiodically, rice is a type of plant

(a) Long day
(b) Short day
(c) Day neutral
(d) None of these

106. The optimum pH of soil for rice cultivation is varies in range

(a) 4 - 6
(b) 5 - 6.5
(c) 6 - 7
(d) 7 - 9

107. IR- 8 variety of rice is introduced in India from

(a) China
(b) Philippines
(c) Indonesia
(d) Taiwan

108. Which of the following variety is a dwarf mutant ?

(a) Jagannath
(b) Jaya
(c) Pusa-2-21
(d) None of these

109. The seed rate of rice cultivation in broadcasting and drilling is sufficient

(a) 100, 60 kg/ha (b) 70, 100 kg/ha
(c) 65, 100 kg/ha (d) 100, 55 kg/ha

110. The dehulled rice grain is also known as

(a) Grey rice (b) Brown rice
(c) Red rice (d) None of these

111. Dalua crop is harvested in

(a) Dec-January (b) July-August
(c) May-June (d) March- April

112. Aman crop is harvested in

(a) Nov-Dec (b) July-August
(c) May-June (d) March-April

113. Aus crop is harvested in

(a) Dec-January (b) July-August
(c) May-June (d) Sept-Oct

114. Which of the following management practices is more problematic in direct seeded rice cultivation

(a) Fertilizer management (b) Weed management
(c) Soil management (d) Water management

115. *Oryza sativa* is having ______ .

(a) 24 chromosomes (b) 41 chromosomes
(c) 22 chromosomes (d) 3 chromosomes

116. *Oryza sativa* is

(a) Diploid (b) Tetraploid
(c) Hexaploid (d) None of these

117. Rice varieties cultivated in India belongs to genus

(a) India (b) Japonica
(c) Javanica (d) None of these

118. Rice varieties cultivated in Japan belongs to genus

(a) Asiatica (b) Japonica
(c) Javanica (d) None of these

119. Optimum temperatures towards maturity of rice lies between

(a) 21-37°C (b) 25-30°C
(c) 20-25°C (d) 15-20°C

120. The centre of origin of rice is

(a) India and Burma (b) China
(c) S. Africa (d) Pakistan

121. The second most cultivated crop of the world is

(a) Rice (b) Wheat
(c) Maize (d) Barley

122. The most cultivated crop in India is

(a) Rice (b) Wheat
(c) Maize (d) Barley

123. In India the productivity of rice is highest in the state

(a) Punjab (b) Haryana
(c) West Bengal (d) Bihar

124. The inflorescence in rice is

(a) Ear (b) Panicle
(c) Spikelets (d) None of these

125. Area under rice cultivation in India is approximately

(a) 39 m ha (b) 45 m ha
(c) 50 m ha (d) None of these

126. In rice _______ is very prominent

(a) Palea (b) Ligule
(c) Lemma (d) None of these

127. Panicle of rice is borne on

(a) Uppermost node (b) Lowermost node
(c) Middle nodes (d) None of these

128. Hull of rice comprised of

(a) Auricle (b) Lemma
(c) Ligule (d) Both (b) and (c)

129. The Dapog method of raising rice nursery was introduced in India from

(a) China (b) S. Africa
(c) USA (d) Philippines

130. Which gene is responsible for dwarfing characters in rice ?

(a) Tift 23 A (b) Dee-gee-woo-gen
(c) Norin - 10 (d) Opaque - 2

131. Golden rice is a rich source of

(a) Vitamin A (b) Vitamin B
(c) Vitamin C (d) Vitamin E

132. Idea of super rice was given by

(a) G. H. Shull (b) G. S. Khush
(c) V. L. Chopra (d) M. S. Swaminathan

133. Dwarf rice is introduced in India in the year

(a) 1974 (b) 1982
(c) 1963 (d) 1954

134. Maximum area under rice is in

(a) India (b) U.S.A
(c) South Africa (d) China

135. "Khaira" disease in rice is due to deficiency of

(a) Zinc (b) Iron
(c) Phosphorus (d) Nitrogen

136. "Akiochi" disease in rice is due to toxicity caused by

(a) Zinc (b) Iron
(c) Phosphorus (d) Potassium

137. Rice variety IR-20 has been evolved at

(a) Phillipines (b) India
(c) Malaysia (d) China

138. Mahsuri rice variety has been introduced in India from

(a) Phillipines (b) China
(c) Malaysia (d) Indonesia

139. Mahsuri an late maturing variety of rice developed by crossing

(a) IR- 20 x IR- 24 (b) Peat x Dee-gee-woo-gen
(c) IR 262 x TKM-6 (d) Taichung 65 x Maying Ebos 80/2

140. IR-20 variety of rice developed by selection from crossing

(a) IR- 20 x IR- 24 (b) Peat x Dee-gee-woo-gen
(c) IR 262 x TKM-6 (d) Taichung 65 x Maying Ebos 80/2

141. The recommended dose of N, P_2O_5 and K_2O kg/ha for dwarf rice are

(a) 100-60-40 (b) 120-40-40
(c) 120-60-40 (d) 100-30-40

142. The shortest duration rice variety is known to nature matures in

(a) 85-90 days (b) 60-75 days
(c) 60 days (d) 70 days

143. Which of the following soils are suitable for rice cultivation

(a) Riverine alluvium, red yellow and red loamy
(b) Hill and submontane soils, tarai soils and laterite soils

(c) Coastal alluvium soils, red sandy soils, mixed red and black soils and medium and shallow black soils
(d) All of these

144. When rice is sown directly in dry field and crop is raised without availability of standing water, this system of cultivation is known as

(a) Dry system (b) Semi-dry system
(c) Boro system (d) Koder system

145. When rice is sown directly in dry field but rain water is compounded when crop is about one and a half to two months old and thereafter enough water is available to the crop, this system of rice cultivation is called

(a) Dry system (b) Semi-dry system
(c) Wet system (d) Boro system

146. Which of the following ratios between nursery area and transplantation is correct when rice seedling are raised by "Dapog method"

(a) 1 : 250 (b) 1: 45
(c) 1: 100 (d) 1: 20

147. In early sixtees Chinese scientists discovered a spontaneous mutant in rice in the Indica variety. It was

(a) Dee-gee-woo-gen (b) Tsai-yuen-chung
(c) Norin - 8 (d) None of these

148. Long duration Indica rice type is

(a) Photo insensitive (b) Photosensitive
(c) Weakly photo sensitive (d) None of these

149. Which is the long duration rice variety ?

(a) Jaya (b) IR-8
(c) Mashuri (d) None of these

150. Which is the most commonly practised method of growing paddy seedling nursery in the northern plains ?

(a) Growing seedlings on raised beds
(b) Growing dry and unsprouted seeds in puddled flat seed beds
(c) Growing rice nursery by DAPOG method
(d) None of these

151. An annual weed commonly found in rice field is

(a) *Allagalliis arvellsis* (b) *Ecllillochloa crusgalli*
(c) *Phalaris minor* (d) None of these

152. Seed rate of transplanted basmati rice is

(a) 25-30 kg/ha (b) 35-40 kg/ha
(c) 45-60 kg/ha (d) 55-70 kg/ha

153. Mat type nursery is related to

(a) Wheat crop (b) Paddy crop
(c) Onion crop (d) Potato crop

154. Zero-till seed drill is used for sowing of

(a) Rice in maize-rice cropping system
(b) Rice in mung rice cropping system
(c) Rice in rice-rice cropping system
(d) All of these

155. The inflorescence in Bajra is

(a) Siliqua (b) Panicle
(c) Raceme (d) Umbel

156. Which one of the following is the centre of origin of Bajra ?

(a) India and Burma (b) China
(c) W. Africa (d) N. Africa

157. In terms of area under cultivation in India bajra ranks

(a) Third (b) Second
(c) Fourth (d) First

158. Optimum temperatures for normal vegetative growth of bajra ranges between

(a) 21-37°C (b) 28-35°C
(c) 20-28°C (d) None of these

159. The recommended seed rate for bajra is ________ .

(a) 6-8 kg/ha (b) 4-5 kg/ha
(c) 14-16 kg/ha (d) 21-24 kg/ha

160. The suitable pH for bajra cultivation is ________ .

(a) 4.0-5.0 (b) 6.0-7.5
(c) 7.5-8.0 (d) 8.5-9.0

161. The optimum plant population for bajra is around

(a) One lacs (b) Two lacs
(c) Four lacs (d) None of these

162. Wherever one irrigation is available, the bajra crop should irrigated at

(a) Late booting to early flowering (b) Tillering stage to early booting
(c) Flowering stage (d) Milking stage

163. Which is the identified critical stage of irrigation in Bajra ?

(a) Tillering (b) Flowering
(c) Both (a) and (b) (d) None of these

164. Which can be transplanted ?

(a) Soybean (b) Bajra
(c) Mustard (d) None of these

165. Which cereal is tillering in nature ?

(a) Soybean (b) Bajra
(c) Sorghum (d) None of these

166. In kera method of bajra sowing the seed is drilled

(a) By hand behind the plough in the furrows
(b) By funnel behind the plough in the furrows
(c) By seed drill
(d) None of these

167. In pora method of bajra sowing the seed is drilled

(a) By hand behind the plough in the furrows
(b) By funnel behind the plough in the furrows
(c) By seed drill
(d) None of these

168. The percentage of the protein content in bajra is around

(a) 12% (b) 8%
(c) 14% (d) 15%

169. Bajra belongs to the family

(a) Cyperaceae (b) Gramineae
(c) Rosaceae (d) Leguminaceae

170. The recommended dose of N, P_2O_5 and K_2O kg/ha for hybrid bajra is

(a) 60-30-40 (b) 100-60-40
(c) 30-20-30 (d) None of these

171. The botanical name of Bajra is

(a) *Pennisetum typhoides* (b) *Pennisetum purpureum*
(c) *Pennisetum spherococcum* (d) *Pennisetum spelta*

172. What is the test weight of bajra ?

(a) 15-18 g (b) 8-12 g
(c) 20-24 g (d) 14-16 g

173. The sowing time of bajra is

(a) March (b) July
(c) December (d) May

174. Bajra crop needs

(a) Warm climate (b) Hot and humid climate
(c) Dry conditions (d) Cold climate

175. The bajra grain is known as

(a) Caryopsis (b) Pod
(c) Kernel (d) None of these

176. The inflorescence of parel millet is usually called

(a) Panicle (b) Siliqua
(c) Umbel (d) Rosette

177. The male-sterile line in bajra was discovered by

(a) G. W. Burton (b) N.E. Borlaug
(c) F. W. Gardner (d) M.S. Swaminathan

178. The male-sterile line was discovered during the year

(a) 1950 (b) 1972
(c) 1982 (d) 1990

179. Barley crop needs

(a) Warm and humid climate (b) Hot and humid climate
(c) Dry and hot climate (d) Cold and dry climate

180. Barley is photoperiodically

(a) Long day plant (b) Short day plant
(c) Day neutral plant (d) None of these

181. The most critical period of irrigation in barley is

(a) Tillering stage (b) CRI stage
(c) Jointing stage (d) Flowering stage

182. What is the recommended seed rate for barley ?

(a) 9-12 kg/ha (b) 15-20 kg/ha
(c) 25-35 kg/ha (d) 75-90 kg/ha

183. The centre of origin of barley is

(a) Asia and Ethiopia (b) America
(c) S. Africa (d) India

184. In India the highest producer of barley is

(a) Punjab (b) U.P.
(c) W. Bengal (d) Haryana

185. The inflorescence in barley is

(a) Siliqua (b) Panicle
(c) Spike (d) Raceme

186. Optimum temperatures towards maturity of barley lies between

(a) 31-37°C (b) 28-30°C
(c) 20-25°C (d) 15-20°C

187. Most optimum temperature range for sowing of barley is

(a) 10-12°C (b) 15-20°C
(c) 12-15°C (d) 25-30°C

188. Which of the following management practices is more problematic in barley cultivation

(a) Fertilizer management (b) Weed management
(c) Water management (d) Soil management

189. Most favourable temperature range for growth of barley is

(a) 28-32°C (b) 20-25°C
(c) 30-37°C (d) 12-15°C

190. Scientific name of barley is

(a) *Hordeum vulgare* (b) *Sorghum bicolar*
(c) *Panicum miliacium* (d) *Argemone maxicana*

191. Which variety of barley is suitable for chapati making ?

(a) Amber (b) RD 137
(c) Neelam (d) Rajkiran

192. The depth of sowing of barley is

(a) 1-2 cm (b) 3-5 cm
(c) 6-8 cm (d) 8-10 cm

193. First irrigation to the barley crop should be given at

(a) CRI stage (b) Tillering stage
(c) Jointing stage (d) Germination stage

194. Which of the following is the ideal temperature lies for tillering stage in barley

(a) 10-15°C (b) 20-25°C
(c) 20-23°C (d) None of these

195. Which of the following variety of barley is suitable for malting ?

(a) Amber (b) RD 137
(c) Rajkiran (d) None of these

196. For control of wild oat in barley it is recommended to apply

(a) Avadex (b) 2, 4-D
(c) Whip super (d) All of these

197. For control of broad leaved weeds in barley it is recommended to apply

(a) Avadex (b) 2, 4-D
(c) Whip super (d) All of these

198. Which is a nematode resistant variety of barley ?

(a) Amber
(b) RD 137
(c) Neelam
(d) Rajkiran

199. Which is a six row variety of barley ?

(a) Amber
(b) RD 137
(c) Rajkiran
(d) None of these

200. Spike of barley is borne on

(a) Uppermost node
(b) Lowermost node
(c) Middle nodes
(d) Centre node

201. Six rowed barley is

(a) *Hordeum vulgare*
(b) *Hordeum distinction*
(c) *Hordeum jubatum*
(d) *Hordeum pusillum*

202. Two rowed barley is

(a) *Hordeum irregulare*
(b) *Hordeum distinction*
(c) Both (a) and (b)
(d) None of these

203. Ratna variety of barley tolerant to saline alkaline conditions released from

(a) Phillipines
(b) Taiwan
(c) Malaysia
(d) IARI

204. Which is a salt tolerant variety of barley ?

(a) Amber
(b) Compana
(c) Conlon
(d) Rajkiran

205. *Hordeum irregulare* is

(a) Two rowed
(b) Six rowed
(c) Four rowed
(d) None of these

206. *Hordeum distinction* is

(a) Two rowed
(b) Six rowed
(c) Four rowed
(d) None of these

207. _______ suggest that saltpeter (KNO_3), and not water, was the "Principle of Vegetation"

(a) J. R. Glauber
(b) Jethro Tull
(c) Robert Boyle
(d) M. S. Swaminathan

208. "Annals of Agriculture" a book was written by

(a) Sir Humphry Davy
(b) Arthur Young
(c) John Woodward
(d) Jethro Tull

209. Who is considered the father of field plot technique

(a) J. B. Boussingault (b) Baker
(c) Arthur Young (d) Liebig

210. In 1943, Agricultural Experiment Station at Rothamsted, England was started by

(a) Milton Whitney (b) Blackman
(c) Lawes and Gilbert (d) Benjamin Franklin

211. Diagnosis and recommendation integrated system (DRIS) is applicable for

(a) Long duration crops (b) Pulse crops
(c) Oilseed crops (d) Vegetables

212. In crop logging, during growing season, plant tissue is sampled every ______ days and analysed for N, sugar,moisture and weight of the young sheath tissue.

(a) 10 (b) 20
(c) 35 (d) 40

213. Botanical name of Flint corn is

(a) *Zea mays amylacea* (b) *Zea mays indurata*
(c) *Zea mays saccharata* (d) None of these

214. Botanical name of sweet corn is

(a) *Zea mays amylacea* (b) *Zea mays indurata*
(c) *Zea mays saccharata* (d) None of these

215. Botanical name of flour corn is

(a) *Zea mays indentata* (b) *Zea mays indurata*
(c) *Zea mays saccharata* (d) *Zea mays amylacea*

216. Botanical name of pop corn is

(a) *Zea mays everta* (b) *Zea mays indurata*
(c) *Zea mays indentata* (d) *Zea mays amylacea*

217. Seminal roots in maize are _______

(a) 30-40 in numbers (b) 3-5 in numbers
(c) 15-20 in numbers (d) 20-25 in numbers

218. Roots help to establish the plant firmly to ground in maize is

(a) Brace roots (b) Seminal roots
(c) Crown roots (d) None of these

219. The female flower of maize is known as

(a) Spadix (b) Tassel
(c) Umbel (d) None of these

220. The male flower of maize is known as

(a) Umbel
(b) Tassel
(c) Panicle
(d) None of these

221. What is the normal seed rate of maize ?

(a) 25 kg/ha
(b) 50 kg/ha
(c) 90 kg/ha
(d) 110 kg/ha

222. The recommended fertilizer dose of N, P_2O_5 and K_2O kg/ha for composite maize are

(a) 80-40-40
(b) 120-30-40
(c) 40-20-30
(d) 60-30-40

223. Botanical name of waxy corn is

(a) *Zea mays indurata*
(b) *Zea mays ceratina*
(c) *Zea mays saccharata*
(d) *Zea mays amylacea*

224. Pop corn is a product of

(a) Maize
(b) Wheat
(c) Bajra
(d) Barley

225. The productivity of maize is highest in which state of the India ?

(a) U.P.
(b) Haryana
(c) Bihar
(d) Punjab

226. Most common herbicide used in maize is

(a) Glyphosate
(b) 2, 4-D
(c) Simazine
(d) Paraquat

227. Who introduced the technique of production of double cross hybrid maize ?

(a) B. P. Pal
(b) D. F. Jones
(c) M. S. Swaminathan
(d) Mendel

228. *Avena sativa* is

(a) Diploid
(b) Tetraploid
(c) Hexaploid
(d) Triploid

229. The centre of origin of sorghum is

(a) India and Africa
(b) America
(c) S. Africa
(d) India

230. Optimum temperatures for growth of sorghum ranges between

(a) 26-30°C
(b) 15-20°C
(c) 20-25°C
(d) 16-20°C

231. The recommended seed rate for sorghum is ________.

(a) 20-25 kg/ha (b) 10-12 kg/ha
(c) 15-20 kg/ha (d) 5-10 kg/ha

232. Sorghum yield is largely reduced mainly due to:

(a) Cold wind at the flowering stage
(b) Weeds
(c) Lack of adequate fertilizer application
(d) Failure of irrigation in flowering stage

233. High yielding hybrid varities of jowr grown under optimal conditions can yield upto

(a) 1000-1500 kg/hectare (b) 2500-3500 kg/hectare
(c) 5000-6000 kg/hectare (d) 1500-2000 kg/hectare

234. Which of the following are not traditional varities of jowar ?

(a) Saoner, Ramkel, Maldanis 35-1
(b) Dagadi, bilichigan , flugar white
(c) Hagari, Yenigar, budhperio
(d) CSH-1,CSH-2, CSH-4, CSV-4, CSV-9, CSV-10

235. Which of the following are improved and high yielding varieties of sorghum

(a) CSH-1,CSH-2, CSV-1,CSV-2, CSV-4
(b) Saoner, Ramkel, Maldandis -2
(c) Dagadi, bilichigan, flugar white
(d) None of these

236. The centre of origin of oat is

(a) Asia minor (b) America
(c) S. Africa (d) India

237. The inflorescence in oat is

(a) Siliqua (b) Panicle
(c) Spike (d) Raceme

238. The optimum temperature range for sowing of oat is

(a) 20-25°C (b) 15-20°C
(c) 10-15°C (d) None of these

239. The recommended seed rate for oat is _______ .

(a) 50-60 kg/ha (b) 12-15 kg/ha
(c) 25-30 kg/ha (d) 70-80 kg/ha

240. The identified critical stage of irrigation in oat is

(a) Tillering stage (b) Flag leaf stage
(c) Flowering stage (d) Maturity stage

241. The most critical stage of irrigation in Oat is

(a) Tillering stage (b) CRI stage
(c) Jointing stage (d) Flowering stage

242. Which of the following composite varieties of maize are very rich in lysine and trytophan essential amino acids

(a) Shakti, Rattan and Protina (b) Vijay and Amber
(c) Sona and Kisan (d) None of these

243. The centre of origin of maize is

(a) India (b) Mexico
(c) S. Africa (d) USA

244. The inflorescence in maize is

(a) Siliqua (b) Panicle
(c) Spikelets (d) Raceme

245. Optimum temperatures towards maturity of maize lies between

(a) 25-37°C (b) 10-15°C
(c) 15-20°C (d) 20-25°C

246. Which of the following is the leading state in maize production ?

(a) Haryana (b) Uttar Pradesh
(c) Punjab (d) Bihar

247. Which of the following is the leading state in rabi maize production ?

(a) Bihar (b) Uttar Pradesh
(c) Haryana (d) Andhra Pradesh

248. Which of the following statement is true in composite maize ?

(a) Seed can be sown year after year with almost same level of performance
(b) Fresh seed is to be obtained for every sowing to get high yields
(c) Seed is a mixture of a few hybrid lines which are mixed before sowing
(d) None of these

249. Male sterile lines are used in India in production of new hybrids of

(a) Rice (b) Maize
(c) Potato (d) Cotton

250. In which crop brace roots are found ?

(a) Bajra (b) Rice
(c) Maize (d) Potato

251. For proper germination of maize temperature requirement is

(a) 18-25°C (b) 25-30°C
(c) 15-18°C (d) 30-35°C

252. For proper grain filling of maize temperature requirement is

(a) 20-25°C (b) 20-33°C
(c) 33-35°C (d) 28-30°C

253. Which vegetable oil is good for heart patient

(a) Peanut oil (b) Mustard oil
(c) Rapeseed oil (d) Sunflower oil

254. In India most commonly followed method of planting sorghum is

(a) Direct seeding in rows
(b) Transplanting on raised seed beds
(c) Direct seedling in puddled beds
(d) None of these

255. Highly tolerant sorghum variety for striga

(a) SPV - IO4 (b) SPV- 222
(c) SAR-2 (d) CSH-5

256. The hybrid seed of CSH - 1 can be produced by crossing

(a) 2077 A x Swarna (b) 2077 A x IS 3691
(c) 2077 A x CS 3541 (d) MSCK 60 A x IS 84

257. The hybrid seed of CSH - 2 can be produced by crossing

(a) MSCK 60-A x CS 351 (b) MSCK 60-A x Swarna
(c) MSCK 60-A x IS 3691 (d) MSCK 60 A x IS 3891

258. CSH is a cross between

(a) 1036 Z x Swarna (b) 2077 A x CS 3541
(c) 1076 A x Swarna (d) 1076 B x Swarna

259. The inflorescence of sorghum is called

(a) Head (b) Rosette
(c) Ear (d) None of these

260. CSH-1 was first sorghum hybrid released in

(a) 1964 (b) 1968
(c) 1967 (d) 1972

261. Botanical name of foxtail millet is

(a) *Coix lachryma* (b) *Echinocloa frumentacea*
(c) *Panicum virgatum* (d) *Paspalum notatum*

262. The recommended seed rate for finger millet is

(a) 8-10 kg/ha (b) 16-20 kg/ha
(c) 12-15 kg/ha (d) 25-30 kg/ha

263. The recommended dose of N, P_2O_5 and K_2O kg/ha for finger millet is

(a) 60-40-30 (b) 120-40-40
(c) 30-20-30 (d) 40-30-40

264. The seed of sawan is called as

(a) Caryopsis (b) Siliqua
(c) Panicle (d) None of these

265. The centre of origin of sawan is

(a) Central Asia (b) India
(c) USA (d) S. Africa

266. The inflorescence in sawan is

(a) Caryopsis (b) Panicle
(c) Spikelets (d) None of these

267. Optimum temperatures towards maturity of sawan ranges between

(a) 21-37°C (b) 25-30°C
(c) 20-25°C (d) None of these

268. Which is a double cross hybrid variety of maize ?

(a) Deccan hybrid (b) Ganga safed-2
(c) Hi starch (d) Paras

269. Which is a three way cross hybrid variety of maize ?

(a) Deccan hybrid (b) Ganga 5
(c) Hi starch (d) Paras

270. Which is a single cross hybrid variety of maize ?

(a) Paras (b) Ganga 5
(c) Ganga 4 (d) Deccan hybrid

271. Which is a high lysine composite variety of maize ?

(a) Shakti (b) Rattan
(c) Protina (d) All of these

272. Which of the following is very early maturing varieties of maize which mature in 65-75 days ?

(a) Sathi (b) Kathri
(c) Teen Pakhia (d) All of these

273. Which of the following maize varieties are classified as "medium early-maturing" type ?

(a) Basi (b) KT 41
(c) Jaunpur, Rudrapur local (d) All of these

274. **Required plant population of jowar may be obtained by using a seed rate of**
(a) 6-9 kg/ha (b) 12-15 kg/ha
(c) 20-25 kg/ha (d) 18-20 kg/ha

275. **The recommended seed rate for proso millet is ______**
(a) 8-10 kg/ha (b) 15-20 kg/ha
(c) 10-15 kg/ha (d) 20-25 kg/ha

276. **The inflorescence in proso millet is**
(a) Caryopsis (b) Panicle
(c) Umbel (d) Cymose

277. **The centre of origin of proso millet is**
(a) India (b) America
(c) S. Africa (d) China

278. **Maximum yield of sawan is obtained at a plant geometry of**
(a) 45 x 20 cm (b) 60 x 30 cm
(c) 20 x 10 cm (d) 105 x 30 cm

279. **The optimum plant population per hectare of sawan is**
(a) 1 lac (b) 2 lacs
(c) 3 lacs (d) 4 lacs

280. **Scientific name of Ragi is**
(a) *Eleusine coracana* (b) *Echinochloa frumentacea*
(c) *Echinocloa crusgalli* (d) *Panicum miliaceum*

281. **The recommended dose of N, P_2O_5 and K_2O kg/ha for sawan are**
(a) 40-30-20 (b) 120-40-40
(c) 40-20-30 (d) 100-50-40

282. **Sawan crop needs**
(a) Warm and humid climate (b) Hot and humid climate
(c) Cold and dry climate (d) None of these

283. **The centre of origin of finger millet is**
(a) India (b) China
(c) USA (d) Europe

284. **Botanical name of finger millet is**
(a) *Elusine coracana* (b) *Echinocloa frumentacea*
(c) *Panicum miliaceum* (d) None of these

285. **Botanical name of cheena is**
(a) *Elusine indica* (b) *Echinocloa crusgalli*
(c) *Panicum miliaceum* (d) None of these

286. Botanical name of kodo is

(a) *Elusine indica* (b) *Echinocloa crusgalli*
(c) *Panicum miliaceum* (d) *Paspalum scrobiculatum*

287. Botanical name of fonio millet is

(a) *Digitaria exilis* (b) *Echinocloa colonium*
(c) *Panicum miliaceum* (d) *Paspalum scrobiculatum*

288. Botanical name of teff millet is

(a) *Setaria italica* (b) *Eragrostis tef*
(c) *Panicum miliaceum* (d) *Paspalum scrobiculatum*

289. Botanical name of foxtail millet is

(a) *Setaria italica* (b) *Echinocloa frumentacea*
(c) *Panicum miliaceum* (d) *Paspalum scrobiculatum*

290. Botanical name of sawan is

(a) *Elusine indica* (b) *Echinocloa frumentacea*
(c) *Panicum miliaceum* (d) *Paspalum scrobiculatum*

291. Spike of oat is borne on

(a) Uppermost node (b) Lowermost node
(c) Middle nodes (d) Centre nodes

292. Which of the following management practices is more problematic in sorghum cultivation

(a) Fertilizer management (b) Weed management
(c) Water management (d) Seed management

293. Sorghum is considered as camel crop because of

(a) Deep root system (b) Resistance to drought
(c) Shallow root system (d) All of these

294. The recommended optimum *kharif* season sorghum seed rate is

(a) 20-25 kg/ha (b) 10-12 kg/ha
(c) 6-9 kg/ha (d) 18-20 kg/ha

295. The seed rate of oat cultivation for seed and fodder, is sufficient

(a) 100, 60 kg/ha (b) 60, 100 kg/ha
(c) 45, 100 kg/ha (d) 100, 45 kg/ha

296. The normal seed rate of Oat is

(a) 60 kg/ha (b) 40 kg/ha
(c) 110 kg/ha (d) 120 kg/ha

297. Scientific name of Oat is

(a) *Hordeum vulgare* (b) *Zea mays*
(c) *Panicum miliacium* (d) *Avena sativa*

298. Grey speck disease in oat it is due to

(a) Manganese (b) Copper
(c) Zinc (d) Magnesium

299. Nitrogen is applied to oat in the proportion of

(a) 50% at basal + 50% at first irrigation
(b) Nil at basal + 50% at tillering + 50% at panicle emergence stage
(c) 50% at basal + 25% at tillering + 25% at panicle emergence stage
(d) None of these

300. The shoot of oat is called

(a) Stem (b) Trunk
(c) Culm (d) Hollow

301. Shoot of oat is

(a) Hollow (b) Solid except nodes
(c) Hollow except at nodes (d) None of these

302. Increase in a yield of a crop a result of increasing a single growth factor is proportional to the decrement from the maximum yield obtainable by increasing the particular growth factor was described by

(a) Ramanujan (b) Bray
(c) Liebig (d) E. A. Mitscherlich

303. Which of the following is not correctly matched

(a) Mustard - White rust
(b) Paddy - Brown rust
(c) Wheat - Red rust
(d) Sugarcane - Red rot

304. J. C. Luthra and his associates developed the solar heat treatment of wheat seeds for the control of

(a) Black rust (b) Loose smut
(c) Brown rust (d) Covered smut

305. Tundu disease (yellow ear rot) of wheat caused by *Clavibacter tritici* is usually associated with

(a) *Heterodera avenae* (b) *Anguina tritici*
(c) *Meloidogyne incognita* (d) *Pratylenchulus* sp.

306. The green ear or downey mildew of pearmillet was first time reported in India by

(a) K. C. Mehta (b) Agrios

(c) E. J. Butler (d) Mundakur

307. Select the disease in which small droplets of pinks or light honey coloured fluid (the honeydew stage) exudes from the spikelets

(a) Ergot of bajra (b) Leaf curl of tomato

(c) Bacterial blight of rice (d) Red rot of sugarcane

308. The management of ergot of bajra may be possible by adopting of

(a) Long crop rotations (b) Keeping seed in 20% salt solution

(c) Spray of oxychloride + zinc (d) All of these

309. The most striking symptom of false smut of rice (*Claviceps oryzae - sativae*) is/are

(a) The colour of kernels finally changes to greenish black

(b) Transformation of individual kernels into large, velvets, green balls

(c) Sometimes the size of kernel are more than twice the diameter of normal grain

(d) All of these

310. Loose smut of wheat is

(a) Internally seed borne (b) Externally seed borne

(c) Both (a) and (b) (d) None of these

311. Covered smut of barley (caused by *Ustilago hordei*) is/are

(a) Internally seed borne (b) Externally seed borne

(c) Air borne (d) Not clearly defined

312. Loose smut of Barley is/are

(a) Internally seed borne (b) Externally seed borne

(c) Soil borne (d) Air borne

313. In rice, which insect causes yellowing and drying of plant leading to condition known as "hopper burn"

(a) Brown plant hopper (b) *Pyrilla*

(c) Gundhi bug (d) None of these

314. Nibbling of rice ear head and defoliation leaving mid rils in rice is caused by

(a) *Hieroglyphus banian* (rice grass hopper)

(b) *Agrotis ipsilon* (cut worm)

(c) *Lyphgma pygmoea* (rice blue beetle)

(d) Both (a) and (b)

Fill in the blanks

1. Central Rice Research Institute is located at ______ .
2. ICARDA is located at ________ .
3. Paddy crop prefers nitrogen in _______ form.
4. Golden rice is rich in _______ vitamin.
5. Indrayani is a scented variety of _____ .
6. Hulling percentage in paddy is _______ .
7. The ratio grain to straw in paddy is ______ .
8. The optimum temperature range for sowing of wheat is ______ .
9. The first irrigation in wheat is normally recommended at ______ .
10. ______ crop is said to be a queen of cereals.
11. Idea of super rice was given by _______ .
12. Dwarf wheat (Mexican) is introduced in India by _____ .
13. Khaira disease in rice is due to the deficiency of ________ .
14. Triticale is a cross between ______ .
15. Protein content of hard wheat is _______ .
16. Casual organism of wheat leaf rust is ________ .
17. Grain to straw ratio in most of the mexican wheat varieties is ________ .
18. Long grain rice is known as ________ .
19. Rice breeding programme in India was started by ______ .
20. Bengal famine first occurred in the year _______ .

True/False

1. *Triticum aestivum* is a hexaploid. **(T)**
2. Khaira disease in rice due to deficiency of nitrogen. **(F)**
3. Rice grain is deficient in lysine. **(T)**
4. Maize belongs to the category bisexual. **(T)**
5. Norin-dwarf gene was introduced in wheat cereal crop. **(T)**
6. Dr. M.S. Swaminathan is pioneer Indian worker in rice. **(F)**
7. The most important food crop of the world is wheat. **(T)**
8. Rice inflorescence is called as panicle. **(T)**
9. Wheat revolution held in India during 1960. **(F)**

10. Country which pioneer in hybrid rice is China. **(F)**
11. Grain used for making beer and whisky is barley. **(T)**
12. Amino acid in barley is Methionine. **(F)**
13. India's position in World Agriculture in terms of wheat third. **(F)**
14. Gas released from paddy field is methane. **(T)**
15. *Oryza sativa* is a diploid species having chromosome 24. **(T)**

Answers Fill in the Blanks

1. Cuttack
2. Syria
3. Ammonical
4. A
5. Paddy
6. 75
7. 2:1
8. 10 – 15° C
9. Crown root initiation stage
10. Wheat
11. G.S. Khus
12. N.E. Borlaug
13. Zinc
14. Wheat x Rye
15. 50–55%
16. *Puccinia recondite*
17. 1:1
18. Basmati
19. Dr. G.P. Hector
20. 1874

ANSWERS

1.(a)	2.(b)	3.(c)	4.(d)	5.(c)	6.(d)	7.(d)
8.(b)	9.(a)	10.(a)	11.(b)	12.(a)	13.(a)	14.(d)
15.(a)	16.(b)	17.(d)	18.(a)	19.(b)	20.(a)	21.(b)
22.(a)	23.(b)	24.(c)	25.(a)	26.(d)	27.(c)	28.(b)
29.(c)	30.(c)	31.(c)	32.(a)	33.(b)	34.(c)	35.(c)
36.(d)	37.(b)	38.(a)	39.(c)	40.(b)	41.(b)	42.(d)
43.(a)	44.(b)	45.(a)	46.(c)	47.(a)	48.(d)	49.(a)
50.(b)	51.(c)	52.(d)	53.(b)	54.(b)	55.(a)	56.(c)
57.(a)	58.(a)	59.(a)	60.(c)	61.(a)	62.(c)	63.(c)
64.(c)	65.(c)	66.(d)	67.(a)	68.(d)	69.(a)	70.(a)
71.(d)	72.(b)	73.(b)	74.(c)	75.(b)	76.(b)	77.(c)
78.(a)	79.(a)	80.(b)	81.(c)	82.(d)	83.(d)	84.(c)
85.(c)	86.(b)	87.(a)	88.(c)	89.(a)	90.(a)	91.(b)
92.(a)	93.(b)	94.(b)	95.(c)	96.(d)	97.(d)	98.(d)
99.(d)	100.(d)	101.(a)	102.(a)	103.(d)	104.(d)	105.(b)
106.(b)	107.(b)	108.(a)	109.(a)	110.(b)	111.(d)	112.(a)
113.(d)	114.(b)	115.(a)	116.(d)	117.(a)	118.(b)	119.(c)
120.(a)	121.(a)	122.(b)	123.(b)	124.(b)	125.(b)	126.(b)
127.(a)	128.(d)	129.(d)	130.(b)	131.(a)	132.(b)	133.(c)
134.(a)	135.(a)	136.(b)	137.(a)	138.(c)	139.(d)	140.(c)
141.(a)	142.(d)	143.(d)	144.(a)	145.(b)	146.(a)	147.(a)
148.(b)	149.(c)	150.(a)	151.(b)	152.(a)	153.(b)	154.(c)
155.(b)	156.(c)	157.(c)	158.(c)	159.(b)	160.(b)	161.(a)
162.(a)	163.(c)	164.(b)	165.(b)	166.(a)	167.(b)	168.(a)
169.(b)	170.(b)	171.(a)	172.(b)	173.(b)	174.(a)	175.(a)
176.(a)	177.(a)	178.(a)	179.(d)	180.(a)	181.(a)	182.(d)
183.(a)	184.(b)	185(c)	186.(b)	187.(b)	188.(b)	189.(d)
190.(a)	191.(c)	192.(b)	193.(b)	194.(a)	195.(b)	196.(a)
197.(b)	198.(d)	199.(a)	200.(a)	201.(a)	202.(c)	203.(d)
204.(a)	205.(a)	206.(a)	207.(a)	208.(b)	209.(a)	210.(c)
211.(a)	212.(c)	213.(b)	214.(c)	215.(d)	216.(a)	217.(b)
218.(a)	219.(a)	220.(b)	221.(a)	222.(a)	223.(b)	224.(a)
225.(d)	226.(c)	227.(b)	228.(c)	229.(a)	230.(a)	231.(b)
232.(b)	233.(c)	234.(d)	235.(a)	236.(b)	237.(b)	238.(b)
239.(a)	240.(c)	241.(d)	242.(a)	243.(b)	244.(c)	245.(d)
246.(c)	247.(a)	248.(a)	249.(b)	250.(c)	251.(a)	252.(a)
253.(d)	254.(a)	255.(d)	256.(c)	257.(b)	258.(b)	259.(a)
260.(a)	261.(a)	262.(a)	263.(a)	264.(a)	265.(a)	266.(b)
267.(c)	268.(a)	269.(c)	270.(a)	271.(d)	272.(d)	273.(d)
274.(b)	275.(a)	276.(b)	277.(a)	278.(c)	279.(d)	280.(a)
281.(a)	282.(a)	283.(a)	284.(a)	285.(c)	286.(d)	287.(a)
288.(b)	289.(a)	290.(b)	291.(a)	292.(b)	293.(b)	294.(b)
295.(b)	296.(a)	297.(d)	298.(a)	299.(a)	300.(c)	301.(c)
302.(d)	303.(c)	304.(b)	305.(b)	306.(c)	307.(a)	308.(d)
309.(d)	310.(d)	311.(b)	312.(a)	313.(a)	314.(a)	

12

Pulse Crop Production

1. **The centre of origin of Pea is**
   (a) Mediterranean (b) USA
   (c) China (d) India
2. **The inflorescence in pea is**
   (a) Axilary receme (b) Umbel
   (c) Panicle (d) Spikelets
3. **In India the maximum area under pea cultivation is in**
   (a) A.P. (b) U.P.
   (c) Rajasthan (d) Tamil Nadu
4. **Optimum temperatures towards flowering and fruiting of pea lies between**
   (a) 21-25°C (b) 25-30°C
   (c) 30-35°C (d) 16-20°C
5. **The recommended seed rate for pea is**
   (a) 60-80 kg /ha (b) 85-110 kg /ha
   (c) 30-35 kg /ha (d) 35-40 kg /ha
6. **Weeds in pea can be controlled by applying**
   (a) Fluchloralin (b) Tribunil
   (c) 2,4-D (d) Both (a) and (b)
7. **Arkel and Asauji are varieties of**
   (a) *Pisum sativum* var hortens (b) *Pisum sativum* var arvense
   (c) *Pisum sativum* var gardens (d) *Cajanus cajan*
8. **Rachna and Harbhajan are varieties of**
   (a) *Pisum sativum* var hortens (b) *Pisum sativum* var arvense
   (c) *Pisum sativum* var gardens (d) *Vigna unguiculata*

9. Pea should be treated with rhizobium inoculation of

(a) *R. japonicum* (b) *R. glycine*
(c) *R. leguminosarum* (d) *R. ciceri*

10. Wilt disease of pea is caused by

(a) *Fusarium oxysporum* (b) *Rhizoctonia solani*
(c) *Erysiphe polygoni* (d) *Uromyces pici*

11. Pea pod borer is

(a) *Phytomyza atricornis* (b) *Etiella zinckenella*
(c) *Septoria pisi* (d) *Fusarium solani*

12. Pea belongs to family

(a) Compositae (b) Malvaceae
(c) Liliaceae (d) Leguminoceae

13. Garden pea is cultivated for its

(a) Beautiful /colourful flowers (b) Green pods used as vegetable
(c) Pasturage for hogs (d) All of these

14. Sweet pea is cultivated for its

(a) Delicious green pods picked up for vegetable and table purposes
(b) Dry seeds for pulse grain
(c) Ornamental purposes alone excellent pasturage for hogs
(d) Pasturage for hogs

15. Field pea is cultivated widely in plains of Uttar Pradesh, Bihar, Ranchi etc. for

(a) Green manuring the poor fields
(b) Obtaining food grains to be used as pulse
(c) Getting peas for canning and freezing purposes
(d) Ornamental flowers and decoration purposes

16. Downy mildew of pea caused by

(a) *Perenospora pisi* (b) *Albugo candida*
(c) *Erysiphae polygani* (d) None of these

17. The symptoms of powdery mildew of pea (Caused by *Erysiphae Polygoni*) first appears on

(a) Stem (b) Leaves
(c) Flowers (d) Roots

18. The flowering time of arhar is

(a) September -November (b) December - February
(c) February - April (d) None of these

19. Which is/are groups of pigeonpea

(a) *C. indius* var. bicolor
(b) *C. indicus* var. flavus
(c) *C. indius* var. odorata
(d) Both (a) and (b)

20. Which group of pigeonpea is having perennial types, late maturing tall and bushy habit ?

(a) *C. indcus* var. flavus
(b) *C. indicus* var. bicolor
(c) *C. indius* var. odorata
(d) None of these

21. Which groups of pigeonpea is having annual types early maturing and smaller plants ?

(a) *C. indicus* var. flavus
(b) *C. indius* var. bicolor
(c) *C. indius* var. odorata
(d) None of these

22. Which group of pigeonpea is commonly known as tur ?

(a) *C. indicus* var. flavus
(b) *C. indius* var. bicolor
(c) *C. indicus* var. odorata
(d) None of these

23. The flowering time of tur is

(a) September - November
(b) December - February
(c) February - April
(d) None of these

24. Branching in pigeonpea begins from __________

(a) 2-4 nodes
(b) 6-10 nodes
(c) 12-16 nodes
(d) None of these

25. The chromosome number of pigeon pea is 2n=

(a) 43
(b) 18
(c) 22
(d) 23

26. What is the optimum row spacing for pigeonpea crop ?

(a) 30-40 cm
(b) 20-35 cm
(c) 40-60 cm
(d) 60-75 cm

27. What is the optimum plant to plant spacing for pigeonpea crop ?

(a) 20 cm
(b) 35 cm
(c) 40 cm
(d) 50 cm

28. Which of these following arhar varieties are medium varieties maturing in 180-200 days ?

(a) Mukta (R-60) Sharda (S-8), BR-183, Dholi 1234, BR-65, Khargone -2
(b) Prabhat, UPAS -120, Pant A-1, Pant A-3
(c) NP (WR 15), T-7, T-17, Hyd. 3A, Hyd. 3C
(d) None of these

29. **Which of the following arhar varieties are late varieties maturing in more than 200 days**

(a) Mukta (R-60) Sharda (S-8), BR-183, Dholi -1234
(b) NP (WR 25), T-7, T-27
(c) T-21, S-5, BS-1
(d) None of these

30. **Which soils are best suited for pigeon pea cultivation ?**

(a) Well drained alluvial and sandy loam
(b) Black cotton soil
(c) Sandy soil
(d) None of these

31. **The insect pod borer is commonly found on**

(a) Sugarcane (b) Rice
(c) Wheat (d) Lentil

32. **Grain of lentil is**

(a) White colour (b) Yellow colour
(c) Brownish colour (d) Red colour

33. **Scientific name of brown lentil is**

(a) *Cicer arietinum* (b) *Cicer kabulium*
(c) *Lens culinaris* (d) *Cicer microphyllum*

34. **Weeds in lentil can be controlled by applying**

(a) Fluchloralin (b) 2,4-D
(c) Ethoxysulfuron (d) Iodosulfuron

35. **Lentil contains about ________ protein**

(a) 18% (b) 15%
(c) 16% (d) 25%

36. **Greengram is originated in**

(a) India (b) China
(c) USA (d) S. Africa

37. **Which one of the following pulse crops is used as a pulse, a fodder and a green manure crop ?**

(a) Groundnut (b) Moong
(c) Soybean (d) Pea

38. **Greengram needs a ________**

(a) Hot and humid climate (b) Cool and humid climate
(c) Hot and dry climate (d) None of these

39. The centre of origin of lentil is

(a) Mediterranean region (b) USA
(c) S. Africa (d) Indonesia

40. The inflorescence in lentil is

(a) Raceme (b) Panicle
(c) Ear (d) Umbel

41. Optimum temperatures for growth of lentil lies between

(a) 21-37°C (b) 18-20°C
(c) 20-25°C (d) 10-15°C

42. Wherever one irrigation is available, the gram crop should irrigated at which stage ?

(a) CRI stage (b) Tillering stage
(c) Pre-flowering stage (d) Flowering stage

43. The dehulled gram grain is known as

(a) Gray gram (b) Brown gram
(c) Red gram (d) None of these

44. Gram crop needs ________ climate

(a) Warm and humid (b) Hot and humid
(c) Dry and hot conditions (d) Cold any dry

45. Photoperiodically, gram is a type of

(a) Long day plant (b) Short day plant
(c) Day neutral plant (d) Intermediate plant

46. Which one of the following pulse crops is used as a pulse, a fodder and a green manure crop ?

(a) Pea (b) Urd
(c) Groundnut (d) None of these

47. Blackgram needs a ________

(a) Hot and humid climate (b) Cool and humid climate
(c) Hot and dry climate (d) None of these

48. The cohesiveness and tenderness of cooked blackgram depends mainly on

(a) The percentage of amylose
(b) The percentage of amylopectin
(c) The proportion of amylose and protein
(d) None of these

49. The centre of origin of gram is

(a) S-W Asia (b) America
(c) S. Africa (d) China

50. **The inflorescence in gram is**
   (a) Axillary raceme (b) Panicle
   (c) Umbel (d) Spikelets

51. **What is the recommended seed rate for blackgram ?**
   (a) 10-15 kg/ha (b) 12-15 kg/ha
   (c) 15-20 kg/ha (d) None of these

52. **Blackgram yield is largely reduced mainly due to :**
   (a) Heavy rains at the flowering stage
   (b) Attack of aphids
   (c) Lack of adequate fertilizer application
   (d) Failure to apply irrigation at the flowering stage

53. **Which of the following is the leading state in blackgram production ?**
   (a) Tamil Nadu (b) Karnataka
   (c) A.P. (d) Assam

54. **The place of origin of blackgram is**
   (a) Burama (b) India
   (c) USA (d) S. Africa

55. **What is the basic aim of nipping in gram ?**
   (a) To divert energy and nutrients from flower head to leaves
   (b) To reduce the plant height
   (c) To protect the plant against disease
   (d) To enlarge branching

56. **Blackgram contains about ________ protein**
   (a) 50% (b) 24%
   (c) 8% (d) 12%

57. **Blackgram contains about ________ carbohydrate**
   (a) 60% (b) 74%
   (c) 12% (d) 46%

58. **Blackgram contains about ________ fat**
   (a) 4.0% (b) 2.6 %
   (c) 6.5 % (d) 1.3%

59. **Blackgram is the richest source of ________ among pulses**
   (a) Lysine (b) Phosphoric acid
   (c) Peptin (d) Methionine

60. **Botanical name of blackgram is**
   (a) *Vigna mungo* (b) *Vigna radiata*
   (c) *Phaseolus radiata* (d) *Pisum sativum*

61. The centre of origin of blackgram is

(a) India (b) China
(c) USA (d) Europe

62. In India the productivity of blackgram is highest in the state

(a) U.P. (b) Bihar
(c) Haryana (d) Andhra Pradesh

63. In India the area of blackgram is highest in the state

(a) Rajasthan (b) Odisha
(c) Karnataka (d) Bihar

64. In India __________State is having maximum production of blackgram

(a) Rajasthan (b) Andhra Pradesh
(c) Odisha (d) Bihar

65. How much seed of gram should be treated with one packet of *Rhizobium* culture

(a) 5 kg (b) 10 kg
(c) 15 kg (d) 20 kg

66. Scientific name of brown gram is

(a) *Cicer arietinum* (b) *Cicer kabulium*
(c) *Cicer gracocum* (d) *Cicer juidaicum*

67. Scientific name of white gram is

(a) *Cicer arietinum* (b) *Cicer kabulium*
(c) *Cicer bijugum* (d) *Cicer anatolicum*

68. Weeds in gram can be controlled by applying

(a) Fluchloralin (b) 2, 4-D
(c) Ethoxysulfuron (d) Iodosulfuron

69. Gram belongs to family

(a) Gramineae (b) Solanaceae
(c) Malvaceae (d) Leguminoceae

70. Which is the leading state in Greengram production ?

(a) Haryana (b) Uttar Pradesh
(c) Maharashtra (d) Tamil Nadu

71. The place of origin of Greengram is

(a) USA (b) India
(c) China (d) S. Africa

72. The fruit of Greengram is known as

(a) Pod (b) Siliquae
(c) Datum (d) None of these

73. The percentage protein content in Greengram seeds is about

(a) 35 (b) 20

(c) 25 (d) 15

74. Pigeonpea generally prefers ________

(a) Cool and dry climate

(b) Warm tropical and sub tropical climate

(c) Temperate humid climate

(d) None of these

75. The recommended seed rate for pigeonpea is

(a) 7-10 kg /ha (b) 12-15 kg /ha

(c) 16-20 kg /ha (d) 21-24 kg /ha

76. The pigeonpea seed contains ______

(a) 10 % protein (b) 28 % protein

(c) 25 % protein (d) 38 % protein

77. Which soils are unfit for pigeon pea cultivation ?

(a) Saline (b) Alkaline

(c) Waterlogged (d) All of these

78. Which is the most commonly practiced method of growing pigeonpea ?

(a) Direct seeding in rows (b) Broadcasting

(c) Transplanting (d) None of these

79. In India the highest acerage of pigeonpea is in the state

(a) Rajasthan (b) Haryana

(c) U.P. (d) Maharashtra

80. In India which state contributes maximum production of pigeonpea ?

(a) Andhra Pradesh (b) Uttar Pradesh

(c) Rajasthan (d) M.P.

81. The inflorescence in pigeonpea is

(a) Axillary raceme (b) Panicle

(c) Umbel (d) Spikelets

82. Scientific name of pigeon pea is

(a) *Vigna radiata [L.]* (b) *Cajanus mungo*

(c) *Cajanus cajan* (d) *Pisum sativum*

83. In India pigeonpea ranks _______ among pulses crops

(a) First (b) Second

(c) Third (d) None of these

84. Grain of field pea is

(a) White colour
(b) Yellow colour
(c) Brownish red colour
(d) Black colour

85. Scientific name of garden pea is

(a) *Pisum sativum* var. hortens
(b) *Pisum sativum* var. arvense
(c) *Pisum sativum* var. gardens
(d) *Cajanus cajan*

86. Scientific name of filed pea is

(a) *Pisum sativum* var. hortens
(b) *Pisum sativum* var. arvense
(c) *Pisum sativum* var. gardens
(d) *Vigna unguiculata*

87. Root rot disease of pea is caused by

(a) *Fusarium solani*
(b) *Rhizoctonia solani*
(c) *Erysiphe graminis*
(d) *Uromyces pici*

88. Pea leaf minor is

(a) *Phytomyza atricornis*
(b) *Etiella zinckenella*
(c) *Erysiphe pisi*
(d) *Alterharia alternata*

89. Rust disease of pea is caused by

(a) *Erysiphe polygoni*
(b) *Ascochyta pisi*
(c) *Uromyces tabae*
(d) *Peronospora viciae*

90. Powdery mildew disease of pea is caused by

(a) *Fusarium oxysporum*
(b) *Peronospora vicial*
(c) *Erysiphe polygoni*
(d) *Pseudomonas pisi*

91. The centre of origin of pigeonpea is

(a) India and Africa
(b) China
(c) USA
(d) Europe

92. Pigeonpea is commonly known as

(a) Arhar
(b) Channa dal
(c) Moong
(d) Matar dal

93. In India pigeonpea is the second most important pulse crop after

(a) Gram
(b) Greengram
(c) Lentil
(d) All of these

94. Which group of pigeon pea is commonly known as arhar ?

(a) *C. indicus* var flavus
(b) *C. indicus* var bicolor
(c) *C. indicus* var odorata
(d) *Cajanal radiata*

95. Pigeonpea has been divided into _________

(a) 6 groups
(b) 2 groups
(c) 4 groups
(d) 8 groups

96. In India leading lentil producing state is

(a) Madhya Pradesh (b) Uttar Pradesh
(c) Rajasthan (d) Tamil Nadu

97. Number of chromosomes in lentil

(a) 2n=14 (b) 2n=24
(c) 2n=34 (d) 2n=46

98. Lentil belongs to family

(a) Compositae (b) Gramineae
(c) Malvaceae (d) Leguminoceae

99. In India the area of Greengram is highest in the state

(a) Punjab (b) Maharashtra
(c) Bihar (d) U.P.

100. In India which state is having maximum production of Greengram ?

(a) Rajasthan (b) Maharashtra
(c) Bihar (d) Madhya Pradesh

101. Greengram contains about __________ protein

(a) 50% (b) 25%
(c) 2.3% (d) 15%

102. Botanical name of Greengram is

(a) *Vigna mungo* (b) *Vigna radiata*
(c) *Phaseolus radiata* (d) *Pisum sativum*

103. The recommended dose of N, P_2O_5 and K_2O kg /ha for *deshi* gram are

(a) 60-40-40 (b) 120-40-40.
(c) 30-60-40 (d) 40-30-40

104. The insect pod borer is commonly found on which crop ?

(a) Gram (b) Barley
(c) Mustard (d) Rice

105. Higher yield of gram could be achieved by

(a) Use of higher dose of phosphate
(b) Closer planting
(c) No nitrogen application
(d) None of these

106. The leading producer of gram is

(a) Burma (b) India
(c) Bangladesh (d) China

107. The country having maximum acreage of gram is

(a) Burma
(b) India
(c) Bangladesh
(d) China

108. Gram is commonly known as

(a) Redgram
(b) Greengram
(c) Bengalgram
(d) Blackgram

109 The fruit of blackgram is known as

(a) Pod
(b) Caryopsis
(c) Siliquae
(d) None of these

110. The protein content in blackgram seeds is about ______

(a) 40%
(b) 50%
(c) 24%
(d) 15%

111. Most commonly followed method of planting blackgram in India is

(a) Direct seeding in rows
(b) Transplanting on raised seed beds
(c) Transplanting in standing water in puddled fields
(d) None of these

112. Blackgram is originated in

(a) India
(b) China
(c) USA
(d) S. Africa

113. Blackgram belongs to family

(a) Liliaceae
(b) Leguminoseae
(c) Compositeae
(d) Malvaceae

114. Crop grown for food, fodder, manuring and soil conservation is

(a) Urd
(b) Groundnut
(c) Soybean
(d) Blackgram

115. Blackgram originated from _______ a wild plant

(a) *Phaseolus sublobatus*
(b) *Phaseolus radiata*
(c) *Vigna radiata*
(d) *Vigna mungo*

116. Early maturing, black coloured bold seeded varieties of blackgram belongs to

(a) *Vigna mungo* var. *niger*
(b) *Vigna mungo* var. *viridis*
(c) *Vigna mungo* var. *helpense*
(d) None of these

117. The inflorescence in blackgram is

(a) Axillary raceme (b) Panicle
(c) Ear (d) Umbel

118. Blackgram is also known as

(a) Mung (b) Urd
(c) Khesari (d) Soybean

119. Higher yield of pea could be achieved by

(a) Use of higher dose of phosphate
(b) Adequate amount of N
(c) No nitrogen application
(d) Closer planting

120. Pea is commonly known as

(a) Arhar (b) Channa
(c) Matar (d) Masoor

121. The ideal temperature for germination of pea is

(a) 22°C to 25°C (b) 15°C to 20°C
(c) 25°C to 30°C (d) None of these

122. The optimum plant population for early pea is around

(a) 2 lacs (b) 3 lacs
(c) 4 lacs (d) 5 lacs

123. The recommended dose of N, P_2O_5 and K_2O kg / ha for garden pea are

(a) 100-40-40 (b) 120-30-40
(c) 20-60-40 (d) 60-30-40

124. Wherever one irrigation is available the pea crop should irrigated at

(a) Pod formation stage (b) Flowering stage
(c) Pre flowering stage (d) Grain filling stage

125. The recommended seed rate for lentil is

(a) 10-15 kg /ha (b) 75-90 kg /ha
(c) 20-25 kg /ha (d) 30-40 kg /ha

126. The leading producer of lentil in the world is

(a) Burma (b) India
(c) Bangladesh (d) China

127. Whole lentil pulse is commonly known as

(a) Arahar (b) Masoor
(c) Matar (d) Malka masoor

128. Which is the largest producer of lentil in the world

(a) India (b) China
(c) USA (d) S. Africa

129. Wherever one irrigation is available the lentil crop should irrigated at

(a) Grain filling (b) Milking stage
(c) Pre-flowering stage (d) Flowering stage

130. In India most commonly followed method of planting Greengram is

(a) Direct seeding in rows
(b) Transplanting on raised seed beds
(c) Direct seeding in puddled beds
(d) None of these

131. The recommended seed rate of Greengram is

(a) 15-20 kg /ha (b) 12-15 kg /ha
(c) 20-25 kg /ha (d) 25-30 kg /ha

132. The inflorescence in Greengram is

(a) Axillary raceme (b) Panicle
(c) Spikelets (d) Umbel

133. Greengram originated from____________ a wild plant

(a) *Vigna radiata* sublobata (b) *Phaseolus radiata*
(c) *Vigna mungo* (d) *Pisum sativum*

134. Which one of the following crop grown for food, fodder, manuring and soil conservation ?

(a) Moong (b) Maize
(c) Cowpea (d) Green gram

135. Greengram belongs to family

(a) Gramineae (b) Leguminoceae
(c) Compositeae (d) Liliaceae

136. Optimum temperatures towards maturity of gram lies between

(a) 21-37°C (b) 25-30°C
(c) 21-24°C (d) 15-20°C

137. The recommended seed rate for gram is

(a) 20-25 kg /ha (b) 75-100 kg /ha
(c) 25-30 kg /ha (d) 50-60 kg /ha

138. For *kharif* season blackgram, the recommended optimum

(a) 15-20 kg /ha (b) 12-15 kg /ha
(c) 10-15 kg /ha (d) 25-30 kg /ha

139. For summer season blackgram, the recommended optimum seed rate is

(a) 6-8 kg /ha (b) 15-20 kg /ha
(c) 30-35 kg /ha (d) 20-30 kg /ha

140. Which is the most important pulse crop in India ?

(a) Chickpea (b) Cowpea
(c) Greenpea (d) None of these

141. What is the recommended seed rate for kabuli gram ?

(a) 20-25 kg /ha (b) 100-125 kg /ha
(c) 30-40 kg /ha (d) 50-60 kg /ha

142. Which is the largest producer of chickpea in the world ?

(a) India (b) China
(c) Nepal (d) USA

143. The ideal temperature for sowing of gram is lies between

(a) 10°C-25°C (b) 5°C-15°C
(c) 25°C-30°C (d) None of these

144. The most important crop grown on conserved moisture in India is

(a) Gram (b) Rice
(c) Barley (d) Oat

145. Awarodhi variety of gram is resistant to

(a) Wilt (b) Flooding
(c) Drought (d) None of these

146. The centre of origin of Greengram is

(a) India (b) S. Africa
(c) USA (d) China

147. In India the productivity of Greengram is highest in the state

(a) U.P. (b) Madhya Pradesh
(c) Tamil Nadu (d) Punjab

148. Application of 20 kgs/ha in pulses increases

(a) Oil content (b) Carbohydrate content
(c) Protein content (d) Fat content

149. Which of the following is the toxic substances of lathyrus (*Lathyrus sativus*) ?

(a) β-N-oxalyl amino alanine (BOAA)
(b) β-N-oxalyl amino alanine (BOAA)
(c) Both (a) and (b)
(d) None of these

150. Continuous consumption of Lathyrus (*Lathyrus sativum*) caused

(a) Anaemia (b) Allergy
(c) Paralysis (d) Night blindness

151. Optimum seed rate of desi Chickpea is

(a) 70-80 kg/ha (b) 90-110 kg/ha
(c) 40-60 kg/ha (d) 100-120 kg/ha

152. Optimum seed rate of kabuli Chickpea is

(a) 30-40 kg/ha (b) 40-70 kg/ha
(c) 80-100 kg/ha (d) 110-130 kg/ha

153. Optimum seed rate of bold seeded lentil is

(a) 20-25 kg/ha (b) 40-60 kg/ha
(c) 60-80 kg/ha (d) 35-40 kg/ha

154. In Plains French bean are generally grown in

(a) Summer season (b) *Kharif* season
(c) *Zaid* season (d) *Rabi* season

155. Optimum seed rate of French bean is

(a) 40-60 kg/ha (b) 100-110 kg /ha
(c) 80-100 kg /ha (d) None of these

156. Optimum seed rate of cowpea for grain purposes is

(a) 30-40 kg/ha (b) 20-25 kg /ha
(c) 30-35 kg /ha (d) 35-40 kg /ha

157. Optimum seed rate of cowpea for green manure purposes is

(a) 35-40 kg /ha (b) 25-30 kg /ha
(c) 40-45 kg /ha (d) 20-25 kg /ha

158. Optimum spacing of field pea is

(a) 25 x 15 cm (b) 30 x 15 cm
(c) 20 x 10 cm (d) 30 x 5 cm

159. Optimum spacing of summer mungbean is

(a) 35 x 20 cm (b) 40 x 20 cm
(c) 30 x 10 cm (d) 35 x 10 cm

160. Germination in urdbean is known as

(a) Epihypogeal type (b) Epigeae type
(c) Hypogeal type (d) None of these

Fill in the Blanks

1. Indian Institute of Pulses Research is situated at _______ .
2. Protein content in soybean seed is ______ .
3. Greengram belong to family ______ .
4. __________is oilseeds as well as pulse crop.
5. Vipula is a variety of __________ .
6. Pulse crops has the highest area in the country is __________ .
7. The blackgram seeds are treated with __________ .
8. *Vigna mungo* is a botanical name of _______ .
9. Malic acid is collected from ________ crop .
10. The centre of origin of greengram is ________ .
11. The inflorescence in greengram is ______ .
12. Blackgram is originated in _______ .
13. Mendel worked on __________.
14. In legumes, red colour of root nodules is attributed to the presence of____ .
15. The term "harvest index" was given by _____
16. CIMMYT is situated in _____
17. Major pulse crop of India is __________
18. The first hybrid pigeonpea released by ICRISAT is __________
19. Centre of origin of cowpea is __________
20. Best suited soil for chickpea cultivation is_____

True/False

1. Leading producer of black gram in India is Andhra Pradesh.**(T)**
2. Cowpea grown for grain, fodder and green manure. **(T)**
3. De-suckering operation is specially followed in chickpea. **(F)**
4. Direct seeding in rows is the most commonly practiced method of growing pigeonpea. **(T)**
5. *Phytomyza atricornis* is pea pod borer. **(F)**
6. Rust disease of pea is caused by *Uromyces fabae*. **(T)**
7. Pigeonpea is commonly known as Arhar. **(T)**
8. The fruit of greengram is known as pod. **(T)**

9. In India the leading state in greengram production is Uttar Pradesh. **(F)**
10. The centre of origin of lentil is S. Africa. **(F)**
11. Direct seeding in rows is the most commonly practiced method of growing Pigeonpea. **(T)**
12. Water logging is extremely harmful to peas. **(T)**
13. Highest area and production of gram is in Punjab. **(F)**
14. Spraying of KCl conserve moisture in chickpea. **(T)**
15. Horsegram is the most ideal crop raised for fodder. **(T)**

Answers Fill in the Blanks

1. Kanpur
2. 40%
3. Leguminaceae
4. Soybean
5. Pigeon Pea
6. Chickpea
7. Rhizobium
8. Blackgram
9. Gram
10. India
11. Axillary raceme
12. India
13. Garden Peas
14. Leg haemoglobin
15. Skoog
16. Maxico
17. Chickpea
18. ICPH-8
19. Africa
20. Black soils

ANSWERS

1.(a)	2.(a)	3.(b)	4.(b)	5.(a)	6.(d)	7.(a)
8.(b)	9.(c)	10.(a)	11.(b)	12.(d)	13.(b)	14.(b)
15.(b)	16.(a)	17.(b)	18.(b)	19.(d)	20.(b)	21.(a)
22.(a)	23.(a)	24.(b)	25.(c)	26.(d)	27.(a)	28.(c)
29.(b)	30.(a)	31.(d)	32.(c)	33.(c)	34.(a)	35.(d)
36.(a)	37.(b)	38.(c)	39.(a)	40.(a)	41.(b)	42.(c)
43.(b)	44.(d)	45.(a)	46.(b)	47.(a)	48.(c)	49.(a)
50.(a)	51.(b)	52.(a)	53.(c)	54.(b)	55.(d)	56.(b)
57.(a)	58.(d)	59.(b)	60.(a)	61.(a)	62.(d)	63.(b)
64.(b)	65.(b)	66.(c)	67.(b)	68.(a)	69.(d)	70.(c)
71.(b)	72.(a)	73.(c)	74.(b)	75.(b)	76.(b)	77.(d)
78.(a)	79.(d)	80.(b)	81.(a)	82.(c)	83.(b)	84.(b)
85.(a)	86.(b)	87.(b)	88.(a)	89.(a)	90.(c)	91.(a)
92.(a)	93.(a)	94.(b)	95.(b)	96.(b)	97.(a)	98.(d)
99.(b)	100.(b)	101.(b)	102.(b)	103.(c)	104.(a)	105.(a)
106.(b)	107.(b)	108.(c)	109.(a)	110.(c)	111.(a)	112.(a)
113.(b)	114.(a)	115.(a)	116.(a)	117.(a)	118.(b)	119.(a)
120.(c)	121.(a)	122.(d)	123.(c)	124.(c)	125.(d)	126.(b)
127.(d)	128.(a)	129.(c)	130.(a)	131.(b)	132.(a)	133.(a)
134.(a)	135.(b)	136.(b)	137.(b)	138.(b)	139.(d)	140.(a)
141.(b)	142.(a)	143.(a)	144.(a)	145.(a)	146.(a)	147.(d)
148.(c)	149.(c)	150.(c)	151.(a)	152.(c)	153.(d)	154.(d)
155.(c)	156.(b)	157.(a)	158.(d)	159.(c)	160.(a)	

13

Oilseed Crop Production

1. **Scientific name of rai is**
   (a) *Brassica juncea* (b) *Eruca sativa*
   (c) *Brassica nigra* (d) None of these
2. **Scientific name of taramira is**
   (a) *Brassica juncea* (b) *Eruca sativa*
   (c) *Brassica nigra* (d) None of these
3. **Scientific name of brown sarson is**
   (a) *Brassica nigra* (b) *Eruca sativa*
   (c) *Brassica campestris* var. dichotoma (d) None of these
4. **Scientific name of Banarsi rai is**
   (a) *Brassica juncea* (b) *Eruca sativa*
   (c) *Brassica nigra* (d) None of these
5. **Oil content in mustard lies between**
   (a) 15-25% (b) 30-40%
   (c) 25-30% (d) 46-60%
6. **Mustard was planted at a plant to plant spacing of 12.5 m with a population of 1,60,000 plants/ha, its row spacing will be**
   (a) 35 cm (b) 50 cm
   (c) 20 cm (d) None of these
7. **Mustard has ______ protein content**
   (a) 15-20% (b) 20-25%
   (c) 25-30% (d) 35-40%

8. Among the oilseed crops mustard ranks ______ in India

(a) First position (b) Second position
(c) Third position (d) None of these

9. Fluchloralin herbicide should be applied in mustard

(a) 1-2 DAS (b) 3-7 DAS
(c) 9-14 DAS (d) None of these

10. Isoproturon herbicides should be applied in mustard at ______ kg ai/ha

(a) 0.5 (b) 0.75
(c) 1.5 (d) None of these

11. Which of the following cruciferous oil seeds is used as a spice in India ?

(a) *Brassica nigra* (Banarasi rai)
(b) *Eruca sativa* (Taramira)
(c) *B. campestris* var. dichotoma (Brown sarson)
(d) None of these

12. Sangam, Br. 36, ITSA and B-54 are varities of

(a) Yellow sarson (b) Brown sarson
(c) White mustard (d) Toria

13. Prakash, Faridkot selection, K-1 and B-85 are the varities of

(a) Yellow sarson (b) White sarson
(c) Raya (d) Toria

14. Test weight of mustard is

(a) 3-5 gram (b) 7-12 gram
(c) 6-8 gram (d) None of these

15. Which of the following countries occupies first position in the world with regard to area and production of sunflower ?

(a) India (b) U.S.S.R (Former)
(c) China (d) S. Africa

16. Maximum area and production of sunflower in India is concentrated in

(a) Gujarat and Andhra Pradesh (b) Tamil Nadu
(c) Maharashtra and Karnataka (d) U.P. & Bihar

17. Which soil types are best suited for cultivation of sunflower ?

(a) Sandy loam soils (b) Loam soils
(c) Black soils with good drainage (d) All these soils

18. The centre of origin of sunflower is

(a) Mexico (b) U.S.A
(c) S. Africa (d) India

19. Which of the following countries occupies first position in the world with regard to area and production of safflower

(a) India (b) U.S.A
(c) China (d) Burma

20. 72 per cent of area and 73% of the production of safflower in India is concentrated in

(a) Gujarat (b) Tamil Nadu
(c) Karnataka (d) Maharashtra

21. Which of the following soil types are best suited for cultivation of safflower ?

(a) Sandy loam soils (b) Loam soils
(c) Heavy soils (d) Black soils

22. Morden is the variety of

(a) Castor (b) Rapeseed
(c) Groundnut (d) Linseed

23. Mustard seed normally contains _______ oil percent

(a) 28 (b) 40
(c) 35 (d) 15

24. Which variety related to oilseeds ?

(a) Ganga-5 (b) Bahar
(c) HD 1982 (d) Varuna

25. Sunflower is sown in the season

(a) *Kharif* (b) *Rabi*
(c) *Zaid* (d) All the season

26. The optimum pH value for sunflower is lies between

(a) 6.5-8.5 (b) 7.0-9.5
(c) 5.0-6.5 (d) None of these

27. The linseed crop besides oil is grown for the extraction of ________

(a) Resins (b) Fibre
(c) Dye (d) Medicine

28. Soybean seed normally contains ______ percent protein

(a) 25 (b) 20
(c) 15 (d) 40

29. The soybean seed normally contains ______ percent oil

(a) 20 (b) 30
(c) 10 (d) 40

30. Which is a bunch type variety of groundnut ?

(a) Punjab- I (b) HD 2189
(c) Narbada 4 (d) TMV-3

31. Which of the following countries occupies first position in the world with regard to area and production of groundnut ?

(a) India (b) USA
(c) China (d) S. Africa

32. Total area under groundnut in India is __________ million hectares

(a) 14.0 (b) 7.5
(c) 16.0 (d) 9.0

33. Total annual production of groundnut in India is about _____ million tones

(a) 15.6 (b) 12.0
(c) 9.0 (d) 5.0

34. The recommended seed rate for bunch type groundnut is _____ .

(a) 80-100 kg/ha (b) 30-40 kg/ha
(c) 50-80 kg/ha (d) None of these

35. Higher yield of groundnut could be achieved by :

(a) Use of higher dose of phosphate
(b) Adequate amount of N
(c) Closer planting
(d) Frequent irrigation

36. Bunch type varieties of groundnut belongs to sub species

(a) Fastigata (b) Javanica
(c) Indica (d) None of these

37. Spreading type varieties of groundnut belongs to sub species

(a) Indica (b) Procumbens
(c) Javanica (d) None of these

38. Which is the largest producer of groundnut in the world ?

(a) India (b) China
(c) USA (d) S. Africa

39. The recommended seed rate for sesamum is _______

(a) 10-12 kg/ ha (b) 6-8 kg/ha
(c) 15-20 kg/ha (d) None of these

40. Sesamum yield is largely reduced mainly due to

(a) Cold wind at the flowering stage
(b) Attack of aphids
(c) Lack of adequate irrigation
(d) None of these

41. Chromosome numbering in *Sesamum indicum* is

(a) 2n = 26 (b) 2n = 32
(c) 2n = 24 (d) None of these

42. Chromosome numbering in *Sesamum prostatum* is

(a) 2n = 26 (b) 2n = 32
(c) 2n = 46 (d) None of these

43. Chromosome numbering in *Sesamum occidentale* is

(a) 2n = 26 (b) 2n = 24
(c) 2n = 28 (d) None of these

44. The centre of origin of castor is

(a) China (b) USA
(c) Africa (d) India

45. In India the productivity of castor is highest in the state

(a) Punjab (b) Haryana
(c) Gujarat (d) Rajasthan

46. In India the area of castor is highest in the state

(a) Rajasthan (b) Haryana
(c) Gujarat (d) Andhra Pradesh

47. In India maximum production of castor is in the state

(a) Punjab (b) Haryana
(c) Gujarat (d) Rajasthan

48. The inflorescence in castor is

(a) Raceme (b) Panicle
(c) Umbel (d) Spikelets

49. Area under castor cultivation in India is approximately

(a) 7 lakh ha (b) 10 lakh ha
(c) 12 lakh ha (d) None of these

50. Optimum temperatures for growth and development of castor lies between

(a) 31-37°C (b) 25-30°C
(c) 20-26°C (d) 15-20°C

51. Sunflower is a ______

(a) Short day crop
(b) Long day crop
(c) Day neutral crop
(d) None of these

52. Which is the largest producer of safflower in the world ?

(a) India
(b) China
(c) U.S.A
(d) S. Africa

53. Safflower contains about _____ per cent of protein

(a) 40
(b) 25
(c) 35
(d) 15

54. The oil content in safflower seeds is about _______ per cent

(a) 25-35
(b) 35-40
(c) 20-25
(d) 15-20

55. Which is the most suitable soil for safflower ?

(a) Sandy and sandy loam
(b) Silty loam and silty clay loam
(c) Clayey soil
(d) Black soil

56. The recommended planting geometry for safflower is _______

(a) 25 x 10 cm
(b) 40 x 15 cm
(c) 45 x 15 cm
(d) 50 x 15 cm

57. Which soil types are best suited for cultivation of niger ?

(a) Sandy loam soils
(b) Loam soils
(c) Clayey loam
(d) All the above

58. Which is a parasitic weed in niger crop

(a) Cascuta
(b) Striga
(c) Orobanche
(d) Dodder

59. Dodder in niger crop can be controlled by pre-emergence application at the rate of 2.0 kg ai per hectare

(a) Tok-E-25
(b) Fluchloralin
(c) Pronamide
(d) None of these

60. The centre of origin of niger is

(a) Ethiopia
(b) U.S.A
(c) China
(d) India

61. A fertilizer dose of 90 kg of nitrogen, 40 kg of phosphorus and 40 kg of potassium per hectare is recommended for

(a) Rainfed sunflower crop
(b) Irrigated linseed crop
(c) Transplanted linseed crop
(d) None of these

62. **Weeds in linseed crop can be controlled by post-emergence application at the rate of 0.5 kg ai per hectare of**
(a) MCPB (b) Glyphosate
(c) 2, 4-D (d) None of these

63. **The centre of origin of linseed is**
(a) India (b) China
(c) S. Africa (d) Mediterranean

64. **The inflorescence in linseed is**
(a) Umbel (b) Cymose
(c) Spikelets (d) Raceme

65. **Optimum temperatures towards seed formation of linseed lies between**
(a) 30-35°C (b) 15-20°C
(c) 20-22°C (d) 20-25°C

66. **The recommended planting geometry for linseed is ______**
(a) 25 x 5 cm (b) 20 x 10 cm
(c) 45 x 15 cm (d) 40 x 15 cm

67. **Linseed oil contains _____ % linolinic acid**
(a) 20-30 (b) 30-40
(c) 40-50 (d) 50-60

68. **The crop cultivated for food, oil, fiber and feed is**
(a) Linseed (b) Groundnut
(c) Soybean (d) Mustard

69. **The linseed fruit is**
(a) Capsule (b) Head
(c) Caryopsis (d) Siliqua

70. **Which oil contains maximum linoleic acid ?**
(a) Safflower (b) Groundnut
(c) Linseed (d) Mustard

71. **The cultivation of which crop requires cool and dry weather ?**
(a) Safflower (b) Potato
(c) Cotton (d) Maize

72. **Safflower oil contains ________ % of linoleic acid**
(a) 55-80 (b) 20-25
(c) 30-40 (d) 40-50

73. Which oil is recommended for patients suffering from cardiac diseases ?

(a) Safflower (b) Groundnut
(c) Linseed (d) Mustard

74. Carthamin is extracted from

(a) Safflower (b) Groundnut
(c) Linseed (d) Mustard

75. The safflower crop besides oil is grown for centuries for extraction of

(a) Kernel (b) Dye
(c) Resins (d) None of these

76. The centre of origin of safflower is

(a) Arabia (b) U.S.A
(c) S. Africa (d) India

77. The inflorescence in safflower is

(a) Head (b) Panicle
(c) Spikelets (d) Umbel

78. Optimum temperatures towards maturity of safflower lies between

(a) 31-37°C (b) 24-32°C
(c) 20-22°C (d) 15-20°C

79. Optimum temperatures for proper germination of safflower is

(a) 31-37°C (b) 24-32°C
(c) 20-22°C (d) 15°C

80. Optimum temperatures for proper growth and development of safflower is

(a) 31-37°C (b) 24-32°C
(c) 20-21°C (d) 15-18°C

81. The inflorescence in sunflower is

(a) Capitulum (b) Panicle
(c) Umbel (d) Raceme

82. Optimum temperatures for sunflower lies between

(a) 31-37°C (b) 28-32°C
(c) 20-22°C (d) 25-27°C

83. The recommended seed rate for sunflower is ______

(a) 10-15 kg/ha (b) 50-60 kg/ha
(c) 20-30 kg/ha (d) 60-70 kg/ha

84. Sunflower contains about _____ per cent of protein

(a) 30 (b) 25
(c) 20 (d) 15

85. The recommended seed rate for castor is _____

(a) 10-15 kg/ha (b) 20-25 kg/ha
(c) 6-12 kg/ha (d) 15-20 kg/ha

86. Castor belongs to ________ family

(a) Euphorbiaceae (b) Cruciferae
(c) Gramineae (d) Malvaceae

87. The usual plant to plant spacing recommended for castor is ______

(a) 20-30 cm (b) 10-20 cm
(c) 60-80 cm (d) 30-50 cm

88. Castor contains about per cent of soil

(a) 40-50 (b) 20-25
(c) 15-20 (d) 25-35

89. Which of the following contains the highest oil content

(a) Castor (b) Soybean
(c) Greengram (d) All of these

90. The recommended dose of N, P_2O_5 and K_2O kg/ha for irrigated mustard are

(a) 100-50-40 (b) 80-40-40
(c) 120-60-40 (d) 60-30-40

91. Which of the following cruciferous oil seeds is known as mustard , in trade

(a) *Brassica campestris* var. sarson and *B. campestris toria*
(b) *Brassica juncea*
(c) *Brassica campestris toria*
(d) *Eruca sativa*

92. Which of the following oil seeds are classified as rapseed in trade ?

(a) *Brassica alba*
(b) *Brassica nigra*
(c) *Brassica campestris* var.sarson , *B. campestris* var. dichotoma, *B. campestris* var. toria and *Eruca sativa*
(d) None of these

93. Which of the following cruciferous oil seeds is not cultivated in India ?

(a) *Brassica alba* (b) *Brassica juncea*
(c) *Brassica nigra* (d) *Eruca sativa*

94. BTHI, M 18 and M 27 are the varities of

(a) Yellow sarson (b) Brown sarson
(c) White sarson (d) Banarasi rai

95. Varuna, Type-1, Type 10, Type 42, Type 151, YS Pb 24, and M3 are the varieties of

(a) Yellow sarson (b) Brown sarson
(c) White sarson (d) Banarasi rai

96. The fruit of mustard is known as

(a) Siliqua (b) Raceme
(c) Caryopsis (d) Ear

97. Optimum temperatures for growth and development of mustard lies between

(a) 30-35°C (b) 25-30°C
(c) 21-25°C (d) 15-20°C

98. The recommended seed rate for mustard is ____ kg/ha

(a) 8-10 (b) 6-8
(c) 15-18 (d) 20-25

99. Higher yield of mustard could be achieved by

(a) Use of higher dose of phosphate
(b) Adequate amount of nitrogen
(c) Adequate amount of sulphur
(d) All of these

100. Mustard yield is largely reduced mainly due to

(a) Cold wind at the flowering stage
(b) Attack of aphids
(c) Lack of adequate fertilizer application
(d) Lack of adequate irrigation

101. Among the oilseed crops sesamum ranks _____ in India

(a) First position (b) Second position
(c) Third position (d) None of these

102. The fruit of sesamum is commonly known to as the

(a) Capsule (b) Pod
(c) Ovate (d) None of these

103. Which is the most suitable soil for sesamum ?

(a) Sandy loam (b) Silty loam
(c) Clayey loam (d) All of these

104. The identified critical stages of irrigation for sesamum is _____

(a) 4-5 leaf stage (b) Flowering stage
(c) Pod development stage (d) All of these stage

105. In case of only one irrigation for sesamum it should be applied at _____

(a) 4-5 leaf stage (b) Reproductive stage
(c) Seed development stage (d) None of these

106. Soybean crop requires N, P_2O_5 and K_2O kg/ha, respectively

(a) 80-30/ 40-20 (b) 90-100, 60-70 and 100-110
(c) 30-60-40 (d) None of these

107. Which of the following countries occupies first position in the world in area and production of sesamum

(a) India (b) USA
(c) China (d) S. Africa

108. Total area under sesamum in India is about

(a) 4.0 million hectares (b) 3.0 million hectares
(c) 7.0 million hectares (d) 5.0 million hectares

109. Total annual production of sesamum in India is about

(a) 16.5 lakh tones (b) 8.0 lakh tones
(c) 5.0 lakh tones (d) 12.4 lakh tones

110. Which of the following soil types are best suited for cultivation of soybean ?

(a) Sandy loam soils (b) Loam soils
(c) Black soils (d) All of these

111. The inflorescence in soybean is

(a) Raceme (b) Panicle
(c) Umbel (d) Spikelets

112. Optimum temperatures towards growth of soybean lies

(a) 31-35°C (b) 26.5-30°C
(c) 20-22°C (d) 16-20°C

113. The recommended seed rate for soybean for grain purpose is _____ kg/ha

(a) 20-30 kg/ha (b) 40-50 kg/ha
(c) 50-60 kg/ha (d) 60-70 kg/ha

114. The recommended seed rate for fodder soybean is _____

(a) 80-100 kg/ha (b) 30-40 kg/ha
(c) 40-50 kg/ha (d) 50-60 kg/ha

115. In groundnut seeds the oil content is about ______ per cent

(a) 45 (b) 25
(c) 30 (d) 15

116. Among the oilseed crops groundnut ranks ____ in India

(a) First position (b) Second position
(c) Third position (d) None of these

117. The gynophore of groundnuts is commonly known as

(a) Peg (b) Pod
(c) Ovate (d) None of these

118. Which is the most suitable soil for groundnut ?

(a) Sandy and sandy loam (b) Silty loam
(c) Clayey loam (d) Loamy clay

119. Which of the soil types are best suited for cultivation of groundnut ?

(a) Sandy loam soils (b) Loam soils
(c) Black soils (d) All these soils

120. Stiff clay soils are not suitable for cultivation of groundnut crop because

(a) Root elongation is hampered (b) Flowering delayed
(c) Pod development is hampered (d) None of these

121. A manure-cum-fertilizer dose of 6.25 tonnes of farmyard manure, 10-20 kg of nitrogen, 20-40 kg of phosphorus and 20-40 kg of potassium per hectare is recommended for

(a) Rainfed groundnut crop (b) Irrigated groundnut crop
(c) Linseed crop (d) Mustard crop

122. Application of 500 kg of gypsum hectare is recommended for growing crop

(a) At sowing time (b) At 20 days before sowing
(c) At pegging stage (d) After irrigation

123. During *kharif* season sunflower takes days to mature

(a) 95-110 days (b) 110-125 days
(c) 120-130 days (d) 80-90 days

124. *Kranti* is a variety of

(a) Toria (b) Linseed
(c) Saron (d) Mustard

125. What is the botanical name of mustard ?

(a) *Brassica juncea* (b) *Brassica compestris*
(c) *Brassica napus* (d) *Brassica rapa*

126. *Ricinus communis* (Castor) is originated in

(a) India (b) Eastern Africa
(c) Tropical America (d) China

127. In Castor nipping is done to

(a) Encourage branching (b) Remove axillary buds
(c) Both (a) and (b) (d) None of these

128. The fruit of castor is

(a) Caryopsis (b) Capsule
(c) Siliqua (d) None of these

129. The turkey red soil is prepared by treating castor oil with

(a) H_2SO_4 (b) HCl
(c) HNO_3 (d) None of these

130. The recommended seed rate for safflower is _______

(a) 50-60 kg/ha (b) 30-40 kg/ha
(c) 25-30 kg/ha (d) 10-15 kg/ha

131. Higher yield of safflower could be achieved by

(a) Use of higher dose of phosphate (b) Adequate amount of nitrogen
(c) Sulphur application (d) All of these

132. Cultivated safflower varieties belongs to species

(a) *Tinctorius* (b) *Palaestinus*
(c) *Oxyacantha* (d) None of these

133. Cultivated safflower has ______ chromosomes

(a) 2n = 18 (b) 2n =24
(c) 2n = 42 (d) 2n = 32

134. Safflower crop needs

(a) Warm and humid climate (b) Hot and humid climate
(c) Dry and hot conditions (d) Cool and dry climate

135. ______ is essential for high oil content and quality of safflower

(a) Calcium (b) Sulphur
(c) Boron (d) Zinc

136. _______ is essential for synthesis of soil and protein

(a) Calcium (b) Sulphur
(c) Boron (d) Zinc

137. Maximum area and production of Niger in India is concentrated in

(a) Gujarat and Andhra Pradesh (b) Tamil Nadu and U.P
(c) Karnataka and Rajasthan (d) M.P., Odisha and Maharashtra

138. The inflorescence in groundnut is

(a) Raceme (b) Umbel
(c) Spikelets (d) None of these

139. Optimum temperatures towards maturity of groundnut lies between

(a) 31-37°C (b) 25-30°C
(c) 20-22°C (d) 18-20°C

140. The recommended seed rate for spreading type groundnut is ______ .

(a) 80-100 kg/ha (b) 40-60 kg/ha
(c) 30-40 kg/ha (d) 60-70 kg/ha

141. The groundnut seed is known as

(a) Kernel (b) Pod
(c) Caryopsis (d) Siliqua

142. The groundnut fruit is known as

(a) Kernel (b) Pod
(c) Caryopsis (d) Siliqua

143. Temperature requirement for proper growth and development of groundnut is

(a) 21-26.5°C (b) 16-20°C
(c) 25 - 30°C (d) 30-35°C

144. Which is essential for root development and proper growth of groundnut ?

(a) Calcium (b) Sulphur
(c) Boron (d) Nitrogen

145. Which is essential for synthesis of oil and protein ?

(a) Calcium (b) Sulphur
(c) Boron (d) Nitrogen

146. The fruit of soybean is commonly known

(a) Peg (b) Pod
(c) Ovate (d) None of these

147. The most suitable soil for soybean is ______

(a) Loam (b) Silty loam
(c) Clayey loam (d) All of these

148. The recommended planting geometry for *kharif* soybean is ________

(a) 30 x 15 cm (b) 20 x 10 cm
(c) 45 x 5 cm (d) 40 x 20 cm

149. The recommended planting geometry for summer soybean is __________

(a) 35 x 10 cm (b) 25 x 15 cm
(c) 30 x 5 cm (d) 40 x 20 cm

150. The recommended depth of sowing for soybean is ________

(a) 4-6 cm (b) 6-7 cm
(c) 3-4 cm (d) 5-6 cm

151. The optimum plant population for soybean is about

(a) One lacs (b) Two lacs
(c) Three lacs (d) None of these

152. The inflorescence in sesamum is

(a) Cymose (b) Panicle
(c) Umbel (d) None of these

153. Optimum temperatures for growth and development of sesamum lies between

(a) 31-37°C (b) 25-27°C
(c) 20-25°C (d) 16-20°C

154. Which is the largest producer of sesamum in the world ?

(a) India (b) USA
(c) S. Africa (d) China

155. Sesamum contains about ____ per cent of protein

(a) 40 (b) 25
(c) 20 (d) None of these

156. The percentage of oil content in sesamum seeds is about ________

(a) 40-50% (b) 15-20%
(c) 10-15% (d) 20-25%

157. The centre of origin of *rai* is

(a) China (b) U.S.A
(c) S. Africa (d) India

158. The centre of origin of brown *sarson* is

(a) U.S.A (b) Afganistan
(c) S. Africa (d) China

159. In India the productivity of mustard is highest in the state

(a) Rajasthan (b) Haryana
(c) Madhya Pradesh (d) U.P.

160. The inflorescence in mustard is

(a) Umbel (b) Receme
(c) Spikelets (d) Siliqua

161. The thinning of mustard is to maintain a plant to plant distance of

(a) 25-30 cm (b) 10-15 cm
(c) 5-10 cm (d) None of these

162. The mustard crop is considered ready for harvesting when its pod turns in

(a) Greenish-yellow colour (b) Yellowish brown colour
(c) Yellowish green (d) None of these

163. The recommended dose of N, P_2O_5 and K_2O kg/ha for rainfed mustard are

(a) 60-20-20 (b) 60-40-40
(c) 40-20-30 (d) 120-30-40

164. Among the oilseed crops castor ranks ______ in India

(a) Eighth position (b) Fifth position
(c) Fourth position (d) Second position

165. The optimum plant population for castor is around

(a) 30000-40000 (b) 100000
(c) 40000-70000 (d) 70000-90000

166. Castor crop needs

(a) Warm and humid climate (b) Hot and humid climate
(c) Dry and hot conditions (d) Cold and dry climate

167. Photoperiodically, castor is a type of

(a) Long day plant (b) Short day plant
(c) Day neutral plant (d) Intermediate plant

168. The recommended dose of N, P_2O_5 and K_2O kg/ha for *deshi* castor are

(a) 60-40-40 (b) 30-40-40
(c) 40-20-30 (d) 120-30-40

169. The recommended dose of N, P_2O_5 and K_2O kg/ha for hybrid castor are

(a) 40-60-40 (b) 120-40-40
(c) 80-60-60 (d) 160-30-40

170. The oil content in sunflower seeds is about ____ per cent

(a) 45 (b) 20
(c) 25 (d) 15

171. The recommended planting geometry for sunflower is _______.

(a) 25 x 10 cm (b) 30 x 15 cm
(c) 45 x 30 cm (d) 40 x 20 cm

172. Sunflower crop needs

(a) Warm and humid climate (b) Hot and humid climate
(c) Dry and hot conditions (d) None of these

173. Sunflower is a _________

(a) Cross pollinated crop (b) Self pollinated crop
(c) Bisexual pollinated crop (d) None of these

174. Groundnut contains about _____ of protein

(a) 50% (b) 30%
(c) 40% (d) 20%

175. Deep ploughing of soil in a groundnut field should be normally avoided because it

(a) Delays harvesting
(b) Delays germination
(c) Makes harvesting difficult
(d) Delays both peg and pod development

176. The recommended planting geometry for groundnut is ________ .

(a) 20 x 10 cm
(b) 40 x 15 cm
(c) 45 x 15 cm
(d) None of these

177. The optimum plant population for groundnut is about

(a) One lacs
(b) Two lacs
(c) Three lacs
(d) None of these

178. The cultivation of which crop requires earthing

(a) Groundnut
(b) Oat
(c) Bajra
(d) Barley

179. Maximum area and production of soybean in India is concentrated in

(a) Andhra Pradesh
(b) Gujarat
(c) Rajasthan
(d) Madhya Pradesh

180. Which of the following countries occupies first position in the world with regard to area and production of soybean ?

(a) S. Africa
(b) America
(c) China
(d) India

181. Pop pod of groundnut is due to the deficiency of

(a) Calcium
(b) Sulphur
(c) Boron
(d) Potassium

182. Soybean crop needs

(a) Warm and moist climate
(b) Hot and humid climate
(c) Dry and hot conditions
(d) None of these

183. Favourable temperature for proper growth and development of soybean is

(a) 26.5-30°C
(b) 28-32°C
(c) 20-25°C
(d) None of these

184. The suitable rhizobium strain for soybean is

(a) Japonicum
(b) Phaseoli
(c) Glycine
(d) None of these

185. The post emergence herbicide for control of broad leaved weeds in soybean is

(a) Isoproturon
(b) Glyphosate
(c) Chlorimuron ethyl
(d) Glymax

186. In India the maximum productivity of sesamum in the state

(a) Odisha (b) Punjab
(c) Rajasthan (d) Maharashtra

187. Which of the following soil types are best suited for cultivation of sesamum ?

(a) Sandy loam soils (b) Loam soils
(c) Black soils (d) All of these

188. Maximum acerage of sesamum in India is in

(a) West Bengal (b) Bihar
(c) Punjab (d) Orissa

189. The centre of origin of sesamum is

(a) India (b) USA
(c) China (d) S. Africa

190. What is the recommended planting geometry for sesame ?

(a) 20 x 10 cm (b) 30 x 10 cm
(c) 35 x 15 cm (d) 40 x 20 cm

191. The optimum plant population for sesamum is around

(a) One lacs (b) Two lacs
(c) Three lacs (d) Four lacs

192. The cultivation of which crop requires thinning ?

(a) Sesamum (b) Barley
(c) Rice (d) Maize

193. The proper sowing time of kharif sesamum is

(a) November - December (b) June - July
(c) March - April (d) None of these

194. Which is the largest producer of rapeseed and mustard in the world ?

(a) S. Africa (b) India
(c) China (d) U.S.A

195. The crops of group rapeseed and mustard belongs to the family _______ .

(a) Cruciferae (b) Composite
(c) Graminae (d) Euphorbiaceae

196. The fruit of rapeseed and mustard is known as ________

(a) Dutum (b) Pod
(c) Siliquae (d) None of these

197. The inflorescence in niger is

(a) Umbel (b) Panicle
(c) Spikelets (d) Capitulum

198. Optimum temperatures for proper growth and development of niger is lies between

(a) 31-37°C (b) 18-23°C
(c) 24-30°C (d) 14-17°C

199. The recommended seed rate for niger is _____

(a) 20-25 kg/ha (b) 5-8 kg/ha
(c) 15-20 kg/ha (d) 35-40 kg/ha

200. The oil content in niger seeds in about _____ per cent

(a) 30-35 (b) 35-40
(c) 25-30 (d) 15-20

201. Optimum temperatures for proper germination of linseed lies between

(a) 31-35°C (b) 25-30°C
(c) 20-25°C (d) 15-20°C

202. The recommended seed rate for line sowing of linseed is ________ .

(a) 25 kg/ha (b) 35 kg/ha
(c) 40 kg/ha (d) 10 kg/ha

203. The recommended seed rate for sowing of linseed by broadcasting is _____

(a) 20-30 kg/ha (b) 35-40 kg/ha
(c) 30-35 kg/ha (d) None of these

204. The recommended seed rate for sowing of utera crop of linseed is ______

(a) 15-20 kg/ha (b) 30-40 kg/ha
(c) 20-25 kg/ha (d) 25-30 kg/ha

205. Linseed is commonly known as ______ in western countries

(a) Coir (b) Flax
(c) Fiber (d) None of these

206. Botanical name of linseed is

(a) *Linum usitatissimum* (b) *Linum angustifolium*
(c) *Linum grandiflorum* (d) *Pisum sativum*

207. Linseed has ________ chromosomes

(a) 2n = 24 (b) 2n = 30
(c) 2n = 32 (d) 2n = 16

208. The most suitable soil for linseed is _______

(a) Loamy clay (b) Silty loam
(c) Clayey soil (d) Black soil

209. The oil content in linseed seeds is about ________

(a) 20-25% (b) 35-45%
(c) 15-20% (d) None of these

210. Linseed contains about _____ per cent of protein

(a) 35 (b) 25
(c) 20 (d) 15

211. The largest producer of linseed in the world is ________

(a) India (b) China
(c) S. Africa (d) USA

212. Embryo of sunflower has

(a) One cotyledone (b) Two cotyledone
(c) Three cotyledone (d) Many cotyledone

213. GAUGI- J 11 GG2, AH 32, Junagarh 113, Punjab- I, AH 334, GAUG-10 are varieties of groundnut recommended for the state

(a) Andhra Pradesh (b) Gujarat
(c) U.P. (d) Maharashtra

214. The centre of origin of groundnut is

(a) Brazil (b) USA
(c) S. Africa (d) India

215. Soybean contains about _____ per cent of protein

(a) 40 (b) 35
(c) 30 (d) 20

216. The oil content in soybean seeds is about ______ per cent

(a) 35 (b) 25
(c) 20 (d) 15

217. Among the oilseed crops soybean ranks ________ in India

(a) First (b) Third
(c) Fourth (d) None of these

218. Niger crop needs

(a) Cool and dry climate (b) Hot and humid climate
(c) Dry and hot conditions (d) All of these

219. Which of the following countries occupies first position in the world with regard to area and production of linseed ?

(a) India (b) USA
(c) China (d) S. Africa

220. The niger crop besides oil is grown for centuries for extraction of

(a) Fiber (b) Dye
(c) Resins (d) All of these

221. Fiber from niger is extracted from

(a) Root (b) Seed
(c) Stem (d) Leaves

222. Niger oil contains _____ % of linolinic acid

(a) 10-15 (b) 20-25
(c) 20-30 (d) 45-55

223. The cultivation of which crop requires cool and dry weather ?

(a) Niger (b) Paddy
(c) Cotton (d) Barley

224. The most suitable soil for niger is _______

(a) Sandy loam and clayey loam (b) Silty loam and silty clay loam
(c) Loamy soil (d) Black soil

225. The recommended planting geometry for niger is ______

(a) 20 x 10 cm (b) 30 x 15 cm
(c) 35 x 15 cm (d) 40 x 20 cm

226. The optimum plant population for niger is around

(a) Two lacs (b) Four lacs
(c) Three lacs (d) Five lacs

227. Which oil contains maximum linolinic acid

(a) Niger (b) Groundnut
(c) Linseed (d) Mustard

Fill in the Blanks

1. Groundnut contains _________ % of oil.
2. Soybean is a _____ plant.
3. The fruit of rapeseed and mustard is known as _______.
4. ________ crop covers maximum area under oilseed in India.
5. Botanical name of Indian mustard is ________ .
6. The office of Director of Oilseeds is located at ______ .
7. The recommended seed rate of sunflower varies from ______ kg/ha.

8. The seed normally required for sowing of sesamum is ______ kg/ha.
9. Castor crop belongs to family ________ .
10. Groundnut crop is originated from ________ .
11. Origin place of soybean is _______.
12. Sunflower crop is also known as an indicator plant for the deficiency of ______.
13. A scale used to describe the growth stages in oilseed rape crop is_______ .
14. The mustard crop is regarded ready for harvest when its pods turn ____ in colour.
15. Linseed is a _______ season crop.
16. Linseed crop belongs to the family ________ .
17. Castor is originated in _______ .
18. Rancidity in sunflower oil is caused by _______.
19. Castor green leaves also fed for ________.
20. In trade *Brassia juncea* is known as _________ .

True/False

1. Sunflower is a cross pollinated crop.**(T)**
2. Varuna is a variety of castor. **(F)**
3. The first sunflower hybrid developed in India is MSFH- 8. **(T)**
4. H4 is a first hybrid of Toria. **(F)**
5. Linseed oil has drying property due to linolenic acid. **(T)**
6. Sebacic acid is prepared from castor. **(T)**
7. Safflower crop is highly resistant to salts. **(T)**
8. Maize is nutrient exhausting crop. **(F)**
9. Interculture in groundnut is avoided at pegging stage. **(T)**
10. The gynophore of groundnut is commonly known as peg. **(T)**
11. The fruit of linseed is known as Seed-ball. **(T)**
12. The most suitable soil for groundnut is loamy clay. **(F)**
13. Kiran is a variety of linseed. **(T)**
14. Niger is a day neutral plant. **(F)**
15. Niger is the associated crop that increase pollination in sunflower. **(T)**

Answers Fill in the Blanks

1. 48%
2. Short day
3. Siliqua
4. Soybean
5. *Brassica compestris*
6. Hyderabad
7. 8 – 10
8. 3 – 4
9. Euphorbiaceae
10. India
11. China
12. Boron
13. Berkankamp scale
14. Yellowish brown
15. Cool
16. Lineaceae
17. Tropical Africa
18. Oxidation
19. Silkworm
20. Mustard

ANSWERS

1.(a)	2.(b)	3.(c)	4.(c)	5.(d)	6.(b)	7.(c)
8.(b)	9.(a)	10.(b)	11.(a)	12.(d)	13.(c)	14.(a)
15.(b)	16.(c)	17.(d)	18.(a)	19.(a)	20.(d)	21.(c)
22.(a)	23.(b)	24.(d)	25.(d)	26.(a)	27.(b)	28.(d)
29.(a)	30.(b)	31.(a)	32.(b)	33.(c)	34.(a)	35.(a)
36.(a)	37.(b)	38.(a)	39.(b)	40.(a)	41.(a)	42.(b)
43.(a)	44.(c)	45.(c)	46.(d)	47.(c)	48.(a)	49.(a)
50.(c)	51.(c)	52.(a)	53.(a)	54.(a)	55.(c)	56.(c)
57.(d)	58.(a)	59.(c)	60.(a)	61.(b)	62.(a)	63.(d)
64.(b)	65.(b)	66.(a)	67.(d)	68.(a)	69.(a)	70.(a)
71.(a)	72.(a)	73.(a)	74.(a)	75.(b)	76.(a)	77.(a)
78.(b)	79.(d)	80.(c)	81.(a)	82.(b)	83.(a)	84.(c)
85.(a)	86.(a)	87.(d)	88.(a)	89.(a)	90.(a)	91.(b)
92.(c)	93.(a)	94.(b)	95.(a)	96.(a)	97.(c)	98.(b)
99.(d)	100.(b)	101.(c)	102.(a)	103.(a)	104.(d)	105.(b)
106.(c)	107.(a)	108.(b)	109.(c)	110.(d)	111.(a)	112.(b)
113.(a)	114.(a)	115.(a)	116.(a)	117.(a)	118.(a)	119.(d)
120.(c)	121.(a)	122.(c)	123.(d)	124.(d)	125.(a)	126.(b)
127.(b)	128.(b)	129.(a)	130.(d)	131.(a)	132.(a)	133.(b)
134.(d)	135.(b)	136.(b)	137.(d)	138.(a)	139.(c)	140.(d)
141.(a)	142.(b)	143.(a)	144.(a)	145.(b)	146.(b)	147.(a)
148.(c)	149.(c)	150.(c)	151.(c)	152.(a)	153.(b)	154.(a)
155.(c)	156.(a)	157.(b)	158.(b)	159.(b)	160.(b)	161.(b)
162.(b)	163.(a)	164.(a)	165.(c)	166.(c)	167.(a)	168.(a)
169.(c)	170.(a)	171.(c)	172.(a)	173.(a)	174.(d)	175.(c)
176.(c)	177.(b)	178.(a)	179.(d)	180.(c)	181.(a)	182.(a)
183.(a)	184.(a)	185(c)	186.(a)	187.(d)	188.(a)	189.(a)
190.(b)	191.(c)	192.(a)	193.(b)	194.(c)	195.(a)	196.(c)
197.(d)	198.(b)	199.(b)	200.(b)	201.(b)	202.(d)	203.(a)
204.(b)	205.(b)	206.(a)	207.(b)	208.(c)	209.(b)	210.(c)
211.(a)	212.(b)	213.(b)	214.(a)	215.(a)	216.(c)	217.(c)
218.(a)	219.(a)	220.(a)	221.(c)	222.(d)	223.(a)	224.(c)
225.(b)	226.(c)	227.(a)				

14

Tuber Crop Production

1. **Potatoes are cultivated by**
   (a) Seeds (b) Foliar buds
   (c) Cutting of roots (d) Buds on tubers
2. **Black heart is a physiological disorder of**
   (a) Tomato (b) Chilli
   (c) Potato (d) Pea
3. **In 1845, the late blight of potato destroyed the potato crop of Ireland was caused by**
   (a) *Phytophtora infestans* (b) *Alternaria solani*
   (c) *Phythium aphanidermatum* (d) *Plasmopara viticola*
4. **The scientific name of potato blackleg pathogen is**
   (a) *Alternaria solani* (b) *Erwinia carotovora*
   (c) *Agrobacterium tumefacieus* (d) *Pseudomonas solanacearum*
5. **Common scab name of potato can be controlled by reducing pH below 5.2 is caused by the pathogen**
   (a) *Streptomyces scabies* (b) *Streptomyces griseus*
   (c) *Erwinia amylovora* (d) *Streptomyces flaveolus*
6. **Potato is a**
   (a) Annual plant (b) Biennial plant
   (c) Perennial plant (d) None of these
7. **Potato as a crop is treated as**
   (a) Annual plant (b) Biennial plant
   (c) Perennial plant (d) None of these

8. **Which is a recommended variety of potato for hills?**
   (a) Kufri alankar (b) Kufri sinduri
   (c) Kufri jyoti (d) Kufri shakti

9. **Which is an extra early variety of potato?**
   (a) Kufri alankar (b) Kufri sinduri
   (c) Kufri red (d) None of these

10. **Which is a late variety of potato for hills?**
    (a) Kufri shakti (b) Kufri sinduri
    (c) Kufri jyoti (d) Kufri alankar

11. **An example of companion cropping is**
    (a) Sugarcane + potato (c) Potato + mustard
    (c) Potato + radish (d) Wheat + pea

12. **Intercropping of mustard with potato is recommended in**
    (a) Replacement series (b) Additive series
    (c) Replacement cum additve series (d) None of these

13. **Which one of the following is TPS variety of potato ?**
    (a) JH 222 (b) HPS 1 / 113
    (c) PJ 376 (d) Jr 5857

14. **The ideal temperature for tuberization in potato is**
    (a) 14°C (b) 18°C
    (c) 24°C (d) 32°C

15. **The main crop of potato is planted from**
    (a) 25 September to 10 October (b) 15 October to 25 October
    (c) 25 October to 25 November (d) 15 March - 25 April

16. **Potato contains _________ of protein**
    (a) 18% (b) 2.0%
    (c) 0.3% (d) 0.6 %

17. **Which is the best soil for potato cultivation ?**
    (a) Sandy loam (b) Clay
    (c) Clay loam (d) Sandy

18. **The weight of potato tuber (for sowing) most economical for higher yields is _______.**
    (a) 20-40 g (b) 30-50 g
    (c) 60-80 g (d) 90-110 g

19. **Which is the most cultivated crop of the world ?**
    (a) Potato (b) Wheat
    (c) Rice (d) Pea

20. The centre of origin of potato is

(a) Peru and Bolivia
(b) Russia
(c) S. Africa
(d) India

21. In India the productivity of potato is highest in the state

(a) Haryana
(b) Tamil Nadu
(c) West Bengal
(d) U. P.

22. Maximum acerage under potato is in the state

(a) Punjab
(b) Tamil Nadu
(c) W. Bengal
(d) U.P

23. Minimum acerage under potato is in the state

(a) Haryana
(b) Tamil Nadu
(c) W. Bengal
(d) Delhi

24. What is the inflorescence in potato ?

(a) Siliqua
(b) Panicle
(c) Spikelets
(d) Umbel

25. Brown rot of potato is caused by

(a) *Alternaria solani*
(b) *Marcophomina phaseolina*
(c) *Pseudomonas solanacearum*
(d) *Rhizotonia solani*

26. _______is a common disease in acidic soils in potato

(a) Early blight
(b) Ring rot
(c) Violet root rot
(d) Scab

27. ________ tilth is preferred for tuberization in potato

(a) Fine
(b) Rough
(c) Cloddy
(d) None of these

28. Which is a resistant variety of potato against late blight and wart ?

(a) Kufri alankar
(b) Kufri sinduri
(c) Kufri jeevan
(d) Kufri shakti

29. Area under potato cultivation in India is approximately

(a) 30 m ha
(b) 1.1 m ha
(c) 40 m ha
(d) 24 m ha

30. Number of cultivated species included under genus *Solonum*

(a) 17
(b) 43
(c) 23
(d) 7

31. Number of wild species included under genus *Solonum*

(a) 111 (b) 117
(c) 154 (d) 9

32. Potato varieties cultivated in India belongs to species

(a) *Andigenum* (b) *Tuberosum*
(c) *Lemmisum* (d) *Solanum*

33. The variety of potato which possesses the resistance to frost conditions is

(a) Kufri chamatkar (b) Kufri alankar
(c) Kufri sheetman (d) None of these

34. The recommended seed rate for potato in northern plains is _______

(a) 5-10 q/ha (b) 15-20 q/ha
(c) 30-35 q/ha (d) 20-25 q/ha

35. Potato belong to the family

(a) Compositeae (b) *Gramineae*
(c) Solanaceae (d) Cruciferae

36. Which growth stage of potato is critical for irrigation ?

(a) Germination (b) Stolon formation
(c) Tuber bulking (d) All of these

37. Most critical weed competition period is up to ______ after planting in potato.

(a) 45-50 days (b) 25-30 days
(c) 35-40 days (d) 30-35 days

38. Depending upon size of the seed tuber the seed rate of potato varies from

(a) 200-700 kg/hectare (b) 1000-1500 kg/hectare
(c) 800-1200 kg/hectare (d) 1500-2000 kg/hectare

39. Potato contains_________ of carbohydrates.

(a) 21% (b) 12%
(c) 3% (d) 6%

40. Potato contains ________ of fat.

(a) 21% (b) 5%
(c) 0.3% (d) 6%

41. Inflorescence in potato is

(a) Racemose (b) Cymose
(c) Panicle (d) Umbel

42. Late blight of potato is caused by

(a) *Alternaria solani*
(b) *Phytophthora infestens*

(c) *Pseudomonas solanacearum*
(d) *Synchutrium endobioticum*

43. The weight of potato tuber (for sowing purpose) most economical for higher yields is ____ .

(a) 15-25 g
(b) 30-50 g
(c) 60-90 g
(d) None of these

44. A potato tuber represents

(a) Enlarged underground stem
(b) Enlarged underground root
(c) Enlarged underground nodules
(d) None of these

45. Potato tuber yield will be ____ , if crop was spaced at 50 x 20c ; and average number of tuber/plant and average weight of tubers are 10 and 50 g, respectively.

(a) 300 q/ha
(b) 500 q/ha
(c) 700 q/ha
(d) 900 q/ha

46. Potato belongs to the family ______

(a) Malvaceae
(b) Solanaceae
(c) Gramineae
(d) Leguminaceae

47. The time of planting of main crop of potato in northern plains is

(a) First fortnight of September
(b) Second fortnight of October
(c) First fortnight of December
(d) Second fortnight of November

48. Which of the following are old traditional potato varieties still cultivated in India ?

(a) Satha, Gola, Phulwa, Up-to-date Darjeeling red round, Craig's defence, Great scot, President
(b) Kufri safed, Kufri kumar, Kufri neela, Kufri sinduri
(c) Kufri chandranmukhi, Kufri khasi-garo, Kufri naveen
(d) None of these

49. Which of the following are not improved modern varieties of potato ?

(a) Kufri red, Kufri kuber, Kufri kisan, Kufri kundan, Kufri chamatkar , Kufri neelmani
(b) Satha, Gola, Phulwa, Up-to-date Darjeeling red round, Carig's defence grat scot, President
(c) Kufri chandramukhi, Kufri khasi-garo, Kufri naveen, Kufri alankar, Kufri jeeven
(d) None of these

50. Most common herbicides used for weed control in potato is/are

(a) Simazine
(b) ToK E-25
(c) Alachlor
(d) All of these

51. Most serious disease of potato is

(a) Early blight of potato (b) Late blight of potato
(c) Stem rot (d) White mold

52. Scientific name of potato is

(a) *Solanum tuberosum* (b) *Solanum melongena*
(c) *Solanum dulcamara* (d) *Solanum americanum*

53. Which is the leading state in production of potato?

(a) U.P. (b) Punjab
(c) M.P. (d) Bihar

54. Optimum plant population ha^{-1} for potato has estimated as

(a) 20,000 - 40,000 (b) 40,000 - 80,000
(c) 50,000 - 70,000 (d) 80,000 - 1,00,000

55. In India, potato crop growing areas are divided into

(a) 7 zones (b) 6 zones
(c) 5 zones (d) 4 zones

56. Tubers exposed to sunlight produce pigment of

(a) Chlorophyll (b) Anthocyanin
(c) Both (a) and (b) (d) None of these

57. Which pest of potato attacks the tuber both in field and storage ?

(a) Potato aphid (b) Lady bird beetle
(c) Potato tuber moth (d) Potato black cut worm

58. What is the optimum temperatures for tuberization of potato ?

(a) 21-37°C (b) 25-30°C
(c) 17-20°C (d) 10-15°C

59. What is the optimum temperatures for growth of potato ?

(a) 14°C (b) 24°C
(c) 25°C (d) 12°C

Fill in the Blanks

1. Tuber of potato is modified ____________
2. Plants having modified rhizomes adopted for storage of foods is called as ____
3. International Potato Centre is located at_________
4. The origin place of potato is _________

5. Potato belongs to the family________
6. When potato is exposed to sun, it develops green colour due to _______
7. Seed potato from hilly regions of India is preferred because _____________
8. Seed rate for potato is _________
9. Seed plot technique of potato is used to produce ________
10. Chipson is a variety of __________________
11. _________ is used to quick up the sprouting of potato________
12. Commonly cultivated potato is _________________ .
13. The variety of potato which possesses the resistance to frost conditions is _________
14. Carbohydrate content in potato______
15. The best method of storing potato at temperature ________

True/False

1. Pushkarnath is a pioneer worker in potato from India. **(T)**
2. Tuber represents enlarged underground stem. **(T)**
3. The best soil for potato cultivation is clay loam. **(F)**
4. Dehauling in potato is done to obtain quality seed tuber. **(T)**
5. The first GM potato developed at CPRI for increasing protein content in tubers consist of genes from Grain amaranthus. **(T)**
6. Potato belongs to the family euphorbiaceae. **(F)**
7. The soil temperature for satisfactory growth of potato tuber should be 17-19° C. **(T)**
8. Seed treatment is done to control the disease. **(T)**
9. Embryo seed dormancy is mainly found in subtropical seeds .**(T)**
10. The major part of agricultural land in India is under cash crop. **(F)**

Answers Fill in the Blanks

1. Stem
2. Tuber
3. Lima
4. South America
5. Solanceae
6. Accumulation of Solanin
7. It is disease free
8. 10 – 15 g/ha
9. Virus free seeds
10. Potato
11. Methyl Bromide
12. *Solanum tuberosum*
13. Kufri shoetman
14. 18 –23%
15. < °C

ANSWERS

1.(d)	2.(c)	3.(a)	4.(b)	5.(a)	6.(c)	7.(a)
8.(c)	9.(a)	10.(b)	11.(a)	12.(a)	13.(d)	14.(b)
15.(c)	16.(b)	17.(a)	18.(b)	19.(a)	20.(a)	21.(b)
22.(d)	23.(d)	24.(b)	25.(c)	26.(d)	27.(a)	28.(c)
29.(b)	30.(d)	31.(c)	32.(b)	33.(c)	34.(b)	35.(c)
36.(d)	37.(a)	38.(d)	39.(a)	40.(c)	41.(a)	42.(b)
43.(b)	44.(a)	45.(b)	46.(b)	47.(b)	48.(a)	49.(b)
50.(d)	51.(a)	52.(a)	53.(a)	54.(d)	55.(c)	56.(c)
57.(c)	58.(c)	59.(b)				

15

Sugar Crop Production

1. **The important by products of sugarcane industry are**

   (a) Bagasse (b) Molasses
   (c) Both (a) and (b) (d) None of these

2. **Bagasse is used as**

   (a) Fuel (b) Fodder
   (c) Fertilizer (d) None of these

3. **Molasses is used for ________**

   (a) Fuel preparation (b) Fodder preparation
   (c) Fertilizer preparation (d) Alcohol preparation

4. **For producing 70 tonnes of normal healthy sugarcane, the crop removes from soil is absent**

   (a) 85-100 kg of nitrogen
   (b) 180-330 of phosphorus
   (c) 60-180 kg of potash and 70-80 kg of calcium
   (d) All of these nutrients

5. **In sugarcane, which leaf is most appropriate for total tissue test**

   (a) The second leaf from the top of the plant
   (b) The third leaf from the top of the plant
   (c) Fully mature leaves midway between the younger centre leave and oldest leaf whorl on the outside
   (d) Fourth fully developed leaf from the top

6. **Crop logging developed by CF Elements is carried out for sugarcane in**

   (a) Brazil (b) Hawai
   (c) New Zealand (d) India

7. **Fertilizers containing Cl^- ions are not suitable for**

(a) Sugarcane (b) Rice
(c) Wheat (d) Potato

8. **In sugarcane, $^{14}CO_2$ is fixed in malic acid. The enzyme is**

(a) RuBP carboxylase
(b) PEP carboxylase
(c) Ribulose phosphate kinase
(d) Fructose phosphatase

9. **Translocation of carbohydrate nutrients usually occur in the form of**

(a) Glucose (b) Maltose
(c) Starch (d) Sucrose

10. **In India the productivity of sugarcane is highest in the state**

(a) Punjab (b) Karnataka
(c) Tamil Nadu (d) None of these

11. **The inflorescence in sugarcane is**

(a) Raceme (b) Open panicle
(c) Spikelets (d) Capitulum

12. **The inflorescence in sugarcane is known as**

(a) Arrow (b) Umbel
(c) Spikelets (d) Spadix

13. **Green tops of sugarcane is known as**

(a) Arrow (b) Head
(c) Spikelets (d) Ear

14. **Optimum temperatures for growth and development of sugarcane ranges between**

(a) 20-25°C (b) 28-32°C
(c) 18-20°C (d) 25-27°C

15. **Essential element for sugar translocation is sugarcane is**

(a) P (b) K
(c) B (d) None of these

16. **Growth stage of sugarcane is critical for irrigation is**

(a) Germination (b) Grand growth phase
(c) Ripening stage (d) None of these

17. **Most critical weed competition period is up to________ after planting in sugarcane.**

(a) 1 Month (b) 4 Months
(c) 3 Months (d) 5 Months

18. In sugarcane most common herbicides used for weed control is /are

(a) Simazine
(b) Atrazine
(c) Alachlor
(d) All of these

19. Which is the most serious disease of sugarcane ?

(a) Red stripe of sugarcane
(b) Red rot of sugarcane
(c) Wilt of sugarcane
(d) Smut of sugarcane

20. How many sets are needed to plant one hectare of sugarcane ?

(a) 20,000-25,000
(b) 35,000-40,000
(c) 40,000-45,000
(d) None of these

21. For ratoon crop of sugarcane it is recommended to grow

(a) Long duration varieties
(b) Short duration varieties
(c) Medium varieties
(d) All of these

22. Juice recovery of sugarcane by ordinary crusher is

(a) 55-65%
(b) 65-75%
(d) 40-50%
(d) All of these

23. Juice recovery of sugarcane by power crusher is

(a) 55-65%
(b) 60-70%
(c) 70-75%
(d) 40-50%

24. Juice recovery of sugarcane by vacuum crusher is

(a) 50-60%
(b) 65-75%
(c) 70-75%
(d) 30-40%

25. Jaggery recovery from sugarcane is

(a) 15-20 %
(b) 9-10%
(c) 10-14%
(d) 25-30%

26. For best autumn crop sugarcane is planted in the month of

(a) August
(b) October
(c) December
(d) March

27. In sugarcane arrowing is

(a) Good, for it increases sugar in the cane
(b) Good, for it increases cross fertilization and seed setting
(c) Considered rather undesirable as it reduces sucrose percent in the cane
(d) None of these

28. The scientific name of sugarcane is

(a) *Saccharum officinarum*
(b) *Saccharum spontaneum*
(c) *Saccharum barberi*
(d) *Saccharum sinense*

29. Sugarcane is native of

(a) New Guinea (b) India
(c) Africa (d) China

30. Which state has highest productivity of sugarcane

(a) Haryana (b) Bihar
(c) Karnataka (d) Tamil Nadu

31. For sugarcane "N" fertilizer application has to be completed within _________ days after planting for better percentage of sugar content in juice

(a) 20 (b) 40
(c) 60 (d) 80

32. The inflorescence in sugarbeat is

(a) Siliqua (b) Open panicle
(c) Spikelets (d) Raceme

33. Optimum temperatures towards germination of sugarbeat lies between

(a) 21.37° C (b) 25-30° C
(c) 20-25° C (d) None of these

34. Optimum temperatures for growth and sugar accumulation in sugarbeat lies between

(a) 21.37° C (b) 25-30° C
(c) 20-22° C (d) 15-18° C

35. The recommended seed rate for sugarbeat is _________

(a) 8-10 kg/ha (b) 16-18 kg/ha
(c) 12-14 kg/ha (d) 22-28 kg/ha

36. Weeds in *Beta vulgaris* can effectively be controlled by.

(a) Pyramin (b) Sirmet
(c) Betanol (d) All of these

37. Weeds in *Beta vulgaris can* effectively be controlled by post emergence application of

(a) Pyramin (b) Sirmet
(c) Betanol (d) All of these

38. Weeds in *Beta vulgaris* can effectively be controlled by pre emergence application of

(a) Pyramin (c) Simazine
(c) Betanol (d) All of these

39. Weeds in *Beta vulgaris* can effectively be controlled by pre emergence application of

(a) Pyramin (b) Sirmet
(c) Both (a) and (b) (d) None of these

40. The seed rate required for sugarbeet is ________

(a) 8-10 kg/ha (b) 12-16 kg/ha
(c) 16-18 kg/ha (d) 20-24 kg/ha

41. Sugarbeet matures in the month of

(a) March-April (b) July- August
(c) October-November (d) January- February

42. For ideal plant population, the seed rate of sugarbeet should

(a) 5-8 kg/ha (b) 8-10 kg/ha
(c) 12-16 kg/ha (d) 28-32 kg/ha

43. Sugarbeet is a

(a) Tuber crop (b) Vegetable crop
(c) Sugar crop (d) None of these

44. Sugarbeet belong to the family ________

(a) Phenopodiaceae (b) Crucifereae
(c) Malvaceae (d) Leguminaceae

45. Sugarbeet belong to the family ________.

(a) Chenopodiaceae (b) Leguminaceae
(c) Liliaceae (d) All of these

46. In sugarbeet the sucrose in found in________

(a) Flower (b) Roots
(c) Leaves (d) All of these

47. Sugarbeet is grown in ________________

(a) *Kharif* season (b) *Rabi* season
(c) *Zaid* season (d) All of these

48. In sugarbeet the sugar content is

(a) 15-16% (b) 17-20%
(c) 6-8% (d) 12-14%

49. In sugarbeet the sugar recovery is ________________

(a) 8-10% (b) 10-12%
(c) 6-8% (d) 18-22%

50. Eksali sugarcane is planted in the month of

(a) July-August
(b) January - February
(c) February- March
(d) November - December

51. The top portion of sugarcane should be selected for seed purposes because bud tissues are rich in

(a) Sucrose (b) Glucose
(c) Lactose (d) Maltose

52. Which element is essential for sugar translocation in sugarcane ?

(a) P (b) K
(c) M (d) Mo

53. Which growth stage of sugarcane is critical for irrigation ?

(a) Germination
(b) Grand growth phase
(c) Ripening stage
(d) None of these

54. Most critical weed competition period is up to ________ months after transplanting exit in sugarcane

(a) 2 (b) 4
(c) 6 (d) 8

55. Sugarcane is a ________

(a) Annual plant (b) Biennial plant
(c) Perennial plant (d) Triennial plant

56. The number of three budded sets needed for planting one hectare of sugarcane is

(a) 15-20 thousands (b) 20-25 thousands
(c) 35-40 thousands (d) 45-55 thousands

57. The lodging of sugarcane is reduced by planting in _______

(a) Flat method
(b) Ridge furrow method
(c) Trench method
(d) All of these

58. Hot water treatment of sugarcane sets is done because

(a) It increases germination
(b) It controls the chloroic streak disease
(c) It promotes the conversion of sucrose into glucose
(d) None of these

59. Method of flat-planting, furrow-planting and trench planting are used in

(a) Sugarcane (b) Potato
(c) Dahlia (d) None of these

60. Sugarcane is a ________

(a) Tropical plant (b) Temperate plant
(c) Sub tropical plant (d) None of these

61. In sugarcane inflorescence the oldest flowers are at

(a) Top (b) Bottom
(c) Middle (d) None of these

62. The sugarcane flowers at the age of ________________

(a) 6-8 months (b) 10-12 months
(c) 14-16 months (d) 7-9 months

63. The soils rich in________ are good for juice quality and lodging resistance

(a) N&P (b) P&Ca
(c) N&K (d) N, P&K

64. The scientific name of noble cane is

(a) *Saccharum officinarum*
(b) *Saccharum spontaneum*
(c) *Saccharum barberi*
(d) *Saccharum sinense*

65. *Saccharum offinarum* is native of

(a) New guinea (b) India
(c) Pakistan (d) China

66. Which state has highest productivity of sugarcane ?

(a) U.P. (b) Bihar
(c) Haryana (d) Tamil Nadu

67. Which state has the largest acreage of sugarcane in country ?

(a) Tamil Nadu (b) Haryana
(c) U.P. (d) Bihar

68. Which state has highest production of sugarcane in country ?

(a) Tamil Nadu (b) Maharashtra
(c) U.P. (d) Haryana

69. Which is the most cultivated sugar crop of the world ?

(a) Sugarcane (b) Sugarbeet
(c) Maize (d) None of these

70. Which is the centre of origin of sugarcane ?

(a) India (b) China
(c) S. Africa (d) Europe

71. Sugarcane belongs to family

(a) Gramineae (b) Chenopodiaceae
(c) Leguminous (d) Liliaceae

72. Sugarcane is a ________.

(a) C_3 plant (b) C_4 plant
(c) Leguminous plant (d) None of these

73. **Sugarcane is a ________.**

(a) Annual plant
(b) Biennial plant
(c) Perennial plant
(d) Triennial plant

74. ***Beta vulgaris* originated in Mediterranean region belongs to**

(a) Gramineae family
(b) Malvaceae family
(c) Chenopodiaceae family
(d) Papavaraceae family

75. **In sugarbeet the source is found in**

(a) Flower
(b) Roots
(c) Leaves
(d) White plant

76. **A good crop of sugarbeet yields ________**

(a) 15-16 t/ha
(b) 20-25 t/ha
(c) 35-50 t/ha
(d) 15-20 t /ha

77. **What is the optimum row to row spacing for sugarbeet ?**

(a) 20-25 cm
(b) 15-20 cm
(c) 40-60 cm
(d) 35-40 cm

78. **What is the optimum plant to plant spacing for sugarbeet ?**

(a) 30-35 cm
(b) 20-25 cm
(c) 10-15 cm
(d) 35-45 cm

79. **What is the optimum plant population for suarbeet ?**

(a) 50,000/ha
(b) 40,000/ha
(c) 60,000/ha
(d) 1,00,000/ha

80. **What is the optimum fertilizer requirement for sugarbeat ?**

(a) 120-80-80
(b) 120-60-40
(c) 120-60-20
(d) None of these

81. **What is the critical period for crop weed competition in sugarbeet ?**

(a) 35-45 days
(b) 50-60 days
(c) 20-25 days
(d) 40-50 days

82. **Most common herbicides used for weed control in sugarcane is/are**

(a) Simazine
(b) Atrazine
(c) Alachlor
(d) All of these

83. **Most serious disease of sugarcane is**

(a) Red stripe of sugarcane
(b) Red rot of sugarcane
(c) Wilt of sugarcane
(d) Smut sugarcane

84. Under high density planting how many sets are needed to plant one hectare of sugarcane

(a) 30,000-38,000 (b) 35,000-45,000
(c) 40,000-48,000 (d) 50,000-55,000

85. Trench method of sugarcane planting is used in

(a) North India (b) Western India
(c) Costal areas (d) All of these

86. With too much increase in nitrogen application the percentage of sugar content in juice is

(a) Decreased (b) Increased
(c) Remain constant (d) None of these

87. *Epricania melanoleuca* is a parasitoid effective against

(a) Sugarcane scale
(b) Blackgram leaf hopper
(c) Sugarcane pyrilla
(d) None of these

88. The soil rich in_________ are good for sugar recovery

(a) P (b) N
(c) Fe (d) K

89. Earthing is done in which of the following crops ?

(a) Sugarcane (b) Rice
(c) Wheat (d) Potato

90. Typing and wrapping in sugarcane is performed to prevent

(a) Insect infestation (b) Lodging
(b) Disease (d) All of these

91. Typing and wrapping in sugarcane is performed in the month

(a) August (b) September
(c) November (d) February

92. Which of the following roots are of permanent type on sugarcane sets ?

(a) Sett roots (b) Shoots roots
(c) Prop roots (d) None of these

93. The inflorescence of sugarcane is known as

(a) Arrow (b) Panicle
(c) Umbel (d) Racemose

94. The favourable temperature for growth of sugarcane lies between

(a) 15-18°C (b) 28-32°C
(c) 19-23°C (d) 33-36°C

95. Adsali sugarcane is planted in the month of

(a) July- August
(b) January-February
(c) March-April
(d) November-December

96. Sugar-beet matures in the month of

(a) May-June
(b) March-April
(c) September-October
(d) January-February

97. The top portion of sugarcane should be selected for seed purposes because bud tissues are rich in

(a) Sucrose
(b) Glucose
(c) Maltose
(d) Fructose

98. The inflorescence of sugarcane is

(a) Caryopsis
(b) Arrow
(c) Ear
(d) Spikelets

99. The average yield of sugarcane in northern India is low compared to southern India is due to

(a) Varietal variability
(b) Soil fertility status
(c) Level of management
(d) Climatic conditions

100. The number of three budded sets needed for planting one hectare of sugarcane is

(a) 15-20 thousand
(b) 20-25 thousand
(c) 35-40 thousand
(d) 45-55 thousand

101. The advantage of planting of sugarcane by trench method ________

(a) Improved tillering
(b) Reduces lodging
(c) Improves germination
(d) Increases the water use efficiency

102. Hot water treatment of sugarcane sets is done because it

(a) Increases germination
(b) Controls diseases
(c) Breaks dormancy
(d) Promotes the conversion of sucrose into glucose

103. The autumn planting of sugarcane is most successful when planted in the month of

(a) July
(b) October
(c) December
(d) March

104. The spring planting of sugarcane is most successful when planted in the month of

(a) December
(b) October
(c) November
(d) February

105. Trench method of sugarcane planting is used mainly in

(a) North India (b) Western India
(c) Coastal areas (d) None of these

106. With too much increase in nitrogen application percentage of sugar content in juice is

(a) Decreased (b) Increased
(c) Remain constant (d) None of these

107. *Epricania melanoleuca* is a parasitoid effective against

(a) Sugarcane leaf hopper
(b) Sugarcane mealy bug
(c) Sugarcane pyrilla
(d) None of these

108. Crop logging is generally used in

(a) Sugarcane (b) Sugarbeet
(c) Maize (d) Potato

109. Sugar recovery from sugarcane juice is

(a) 15-20% (b) 7-9%
(c) 6-10% (d) 12-14%

110. Sucrose content in sugarcane is

(a) 13-24% (b) 8-12%
(d) 25-35% (d) None of these

111. "Arrowing" is

(a) Rooting in sugarcane
(b) Stem elongation in sugarcane
(c) Sugar accumulation in sugarcane
(d) Flowering of sugarcane

112. Which portion of sugarcane stalk is not preferred for sowing purposes ?

(a) Top one-third to half portion
(b) Top one-fourth to half portion
(c) Bottom one-fourth portion
(d) Bottom one-third portion

113. For optimum population sugarcane seed sets should essentially have __________

(a) Four buds (b) Three buds
(c) Two buds (d) Five buds

114. Which state has the largest acreage and highest production of sugarcane in country ?

(a) Tamil Nadu (b) Karnataka
(c) U.P. (d) Punjab

115. Which of the following roots are of permanent type on sugarcane sets ?

(a) Sett roots
(b) Shoot roots
(c) Prop roots
(d) None of these

116. The emergence of inflorescence in sugarcane is

(a) Arrowing
(b) Furrowing
(c) Nipping
(d) None of these

117. The best suited temperature for growth of sugarcane lies between

(a) 15-20°C
(b) 28-32°C
(c) 20-25°C
(d) 25-27°C

118. For ideal plant population, the seed rate of sugarbeet should be

(a) 5-7 kg/ha
(b) 8-10 kg/ha
(c) 15-20 kg/ha
(d) 28-32 kg/ha

119. Higher yield of sugarbeat could be achieved by:

(a) Use of higher dose of phosphate
(b) Adequate amount of nitrogen
(c) No nitrogen application
(d) Use of higher dose of potassium

120. Problem in sugar extraction from roots is due to

(a) Use of higher dose of phosphate
(b) Use of excess dose of N
(c) No nitrogen application
(d) Use of excess dose of potassium

121. The first man to establish sugar factory by processing of beet roots

(a) F.C. Achard
(b) B.P. Pal
(c) N.E. Borlouge
(d) K.C. Mehta

122. Sugarbeet is a

(a) Salad crop
(b) Tuber crop
(c) Sugar crop
(d) None of these

123. 'Arrowing' is known as

(a) Flower primodia initiation in wheat
(b) Flowering of cotton
(c) Flowering of wheat
(d) Flowering of sugarcane

124. Portion of sugarcane stalk is preferred for sowing purposes is

(a) Top one-third to half portion
(b) Top one-fourth portion
(c) Bottom one-fourth portion
(d) None of these

125. Sugarcane seed sets should essentially have ________

(a) Four buds
(b) Three buds
(c) Two buds
(d) None of these

126. For harvest of sugarcane the brix ratio of the lower and above portion of the stem should be

(a) 0.9-1.0 (b) 0.4-0.6
(c) 0.6-0.8 (d) None of these

127. Select the wrong pair

(a) Red rot of sugarcane : *Colletotrichum gloeosporides*
(b) Ergot of bajra : *Claviceps fusiformis*
(c) Erogt of rye : *C. purpurea*
(d) False smut of rice : *C. oryzae*

128. The false smut of sugarcane can be controlled by adopting

(a) Avoid the practice of ratooning
(b) Disinfection of salts before planting
(d) Removal of smuted whips from the field
(d) All of these

129. Sugarcane root borer damages

(a) Leaves (b) Stem
(c) Roots (d) None of these

130. The bushy appearance with dead heart in sugarcane at 6th inter-node is due to

(a) Excess irrigation (b) Top borer
(c) Excess nitrogen (d) Mealy bug

131. Stripping of sugarcane leaves minimizes the attack of

(a) Mealy bugs (b) Pyrilla
(c) Top shoot borer (d) None of these

132. Sugarcane leaf hopper (Pyrilla perpusilla) is controlled by release of lepidopteran parasitoid

(a) *Epiricania melanoleuca*
(b) *Granulosis virus*
(c) *Sturmios inferens*
(d) *Trichogramma chilonis*

133. *Beta vulgaris* originated in Mediterranean region belongs to family

(a) Gramineae (b) Malvaceae
(c) Chenopodiaceae (d) Papaveraceae

134. What is the ideal temperature for tuberization in potato

(a) 10°C (b) 18°C
(c) 20°C (d) 30°C

Fill in the Blanks

1. Sugarcane belongs to _________________ family.
2. Sugarcane is harvested at _________ brix reading.
3. Highest sugarcane production state in India is _________
4. Recommended planting time of pre-seasonal sugarcane is _______ .
5. The duration of adsali sugarcane is _________ months
6. The inflorescence of sugarcane is known as ______ .
7. *Saccharum officinarum* is a native of ______ .
8. The best suited temperature for growth of sugarcane lies between _____ .
9. AICRP on sugarcane was started in ______ .
10. The top portion of sugarcane should be selected for seed purposes because bud tissues are rich in _________ .
11. Essential element for sugar translocation in sugarcane is _________.
12. Most serious disease of sugarcane is_____________.
13. Indian Institute of Sugarcane Research is situated at_________
14. The first man to establish sugar factory by processing of beetroots ________ .
15. Sugarcane belong to the family____________
16. Sugarbeet is grown in _________ season .
17. The seed rate required for sugarbeet is _________kg/ha.
18. In sugarbeet the sugar content is_________ % .
19. In sugarbeet the sucrose is found in___________
20. Sugarcane is a ____________ plant.

True/False

1. *Epricania melanoleuca* is a parasitoid effective against sugarcane pyrilla.**(T)**
2. Smut is the most serious disease of sugarcane. **(F)**
3. Sugarcane is a C_3 plant. **(F)**
4. Sugarcane belongs to family Gramineae. **(T)**
5. The most cultivated sugar crop of the world is sugarbeet. **(F)**

Answers Fill in the Blanks

1. Gramineae
2. 20
3. Maharashtra
4. October - November
5. 18
6. Arrow
7. New Guinea
8. 26-32° C
9. 1960
10. Galactose
11. K
12. Red rot
13. Lucknow
14. F.C. Achard
15. Chenopodiaceae
16. *Rabi*
17. 8 – 10
18. 15 - 16
19. Roots
20. Perennial

ANSWERS

1.(c)	2.(a)	3.(d)	4.(d)	5.(d)	6.(b)	7.(a)
8.(b)	9.(d)	10.(d)	11.(b)	12.(a)	13.(a)	14.(b)
15.(a)	16.(b)	17.(b)	18.(d)	19.(b)	20.(b)	21.(b)
22.(a)	23.(b)	24.(c)	25.(b)	26.(b)	27.(c)	28.(a)
29.(b)	30.(d)	31.(c)	32.(b)	33.(d)	34.(c)	35.(a)
36.(a)	37.(c)	38.(a)	39.(c)	40.(a)	41.(d)	42.(b)
43.(a)	44.(a)	45.(a)	46.(b)	47.(b)	48.(a)	49.(b)
50.(b)	51.(b)	52.(a)	53.(b)	54.(b)	55.(c)	56.(c)
57.(c)	58.(b)	59.(a)	60.(a)	61.(b)	62.(b)	63.(b)
64.(a)	65.(b)	66.(d)	67.(c)	68.(c)	69.(a)	70.(a)
71.(a)	72.(b)	73.(c)	74.(b)	75.(b)	76.(c)	77.(c)
78.(b)	79.(d)	80.(a)	81.(a)	82.(c)	83.(b)	84.(d)
85.(b)	86.(a)	87.(c)	88.(c)	89.(a)	90.(b)	91.(a)
92.(b)	93.(a)	94.(b)	95.(a)	96.(b)	97.(b)	98.(b)
99.(d)	100.(c)	101.(b)	102.(d)	103.(b)	104.(d)	105.(c)
106.(a)	107.(c)	108.(a)	109.(c)	110.(a)	111.(d)	112.(c)
113.(b)	114.(c)	115.(b)	116.(a)	117.(b)	118.(b)	119.(b)
120.(b)	121.(a)	122.(c)	123.(d)	124.(a)	125.(b)	126.(a)
127.(a)	128.(d)	129.(b)	130.(a)	131.(a)	132.(a)	133.(c)
134.(b)						

16

Jute and Allied Fibres

1. **The inflorescence in cotton is**

(a) Axillary (b) Corymb
(c) Spikelets (d) Umbel

2. **In India the production of cotton is highest in the state**

(a) Bihar (b) U.P
(c) Gujarat (d) Maharashtra

3. **The centre of origin of cotton is**

(a) India (b) USA
(c) S. Africa (d) China

4. **Maximum area under cotton cultivation in India is in**

(a) U.P. (b) Bihar
(c) Punjab (d) Maharashtra

5. **Which part of cotton plant is used for tissue test ?**

(a) All the above ground portion
(b) The youngest fully mature leaves of the main stem
(c) Fourth fully developed leaf from the top
(d) The four upper leaf blades

6. **Select the wrong pair**

(a) Black arm of cotton: *X. campestris pv. malvacearum*
(b) Bacterian leaf blight of rice: *Xanthomonas compestris pv. oryzae*
(c) Red stripe of sugarcane : *Pseudomonas rubrilineaus*
(d) Ergot of bajra : *Erwinia carolovora*

7. Curling of leaf upwards and yellowing of terminal cotton shoots is characteristic symptom of presence of

(a) Cotton Aphid (*Aphis gossypii*)
(b) Cotton leaf hopper (*Amrasca devastans*)
(c) Pink boll worm (*Pectinophora gossypiella*)
(d) None of these

8. Which bacteria is responsible for staining of the cotton fibre

(a) *Mycobacterium* sp. (b) *Clostridium botulinum*
(c) *Nematospora gossypii* (d) *Bacillus* sp.

9. In cotton, the serious sucking pest is

(a) Leaf hopper (b) Spotted boll worm
(c) Pink boll worm (d) American boll worm

10. In India the productivity of cotton is highest in the state

(a) Punjab (b) Rajasthan
(c) Bihar (d) U.P.

11. Heavy shedding of buds and bolls occurs in cotton due to

(a) Competitive relationship
(b) Deficiency of phosphorus in the soil
(c) Deficiency of magnesium in the soil
(d) Water stress at bud formation stage

12. Suitable herbicides for weed control in cotton crop is

(a) Diuron (b) Trifluralin
(c) Fluchloralin (d) All of these

13. Topping in cotton crop should be done at________

(a) 80-90 days stage (b) 50-60 days stage
(c) 40-50 days stage (d) 60-70 days stage

14. Topping in cotton is done to

(a) Enhance flowering (b) Enhance sympodial branching
(c) Enhance terminal growth (d) Enhance seed size

15. Topping in cotton is done at ________ height

(a) 1.0-1.20 m (b) 1.5-2.0 m
(c) 1.0-1.5 m (d) 2.0-3.0 m

16. The fruit in cotton is known as

(a) Boll (b) Square
(c) Capsule (d) Mareck

17. **Which one of the following crops has the highest consumption of pesticide ?**
    (a) Maize (b) Cotton
    (c) Oilseeds (d) Wheat
18. **Cotton is originated in**
    (a) India (b) Tropical America
    (c) China (d) Indonesia
19. **The oil content in the cotton seed lies between**
    (a) 40-50% (b) 15-25%
    (c) 30-40% (d) 10-15%
20. **Which of the following cultivated species of cotton grown in India covers half of the cotton area of the country ?**
    (a) *Gossypium hirsutum* (b) *Gossypium lobatum*
    (c) *Gossypium arboreum* (d) *Gossypium tomenstosum*
21. **Cotton is cultivated in India on**
    (a) 4.0 million hectares (b) 2.9 million hectares
    (c) 8.0 million hectares (d) 7.0 million hectares
22. **The first commercial Hybrid cotton H-4 was developed by**
    (a) C.T. Patel (b) H. Katarki
    (c) R.S. Singh (d) B.P. Pal
23. **The first interspecific hybrid cotton Varalakshmi was developed by**
    (a) R.S. Singh (b) H. Katarki
    (c) J.C. Mehta (d) M.S. Swaminathan
24. **DCH-32, an extra long staple hybrid cotton was released in**
    (a) 1970 (b) 1975
    (c) 1981 (d) 1995
25. **DCH-32, an extra long staple hybrid cotton was developed via**
    (a) Heterosis breeding (b) Inter-specific hybridization
    (c) Mutation (d) Selection method
26. **The first hybrid cotton in the world was developed at**
    (a) Surat (b) Varanasi
    (c) Nagpur (d) Lucknow
27. **First interspecific hybrid cotton was released is**
    (a) U.P (b) Gujarat
    (c) Punjab (d) Haryana

28. Reproductive structures in succession in cotton

(a) Square-bolls-flowers
(b) Squares-bolls-flowers
(c) Square-flowers-bolls
(d) None of these

29. Egyptian cotton is

(a) *Gossypium lobatum*
(b) *Gossypium tomenstosum*
(c) *Gossypium barbadense*
(d) *Gossypium davidsonii*

30. Number of species included under genus gossypium is

(a) 20
(b) 26
(c) 24
(d) 28

31. Area under cotton cultivation India ranks

(a) First
(b) Second
(c) Third
(d) None of these

32. In production of cotton India ranks

(a) First
(b) Second
(c) Third
(d) Fourth

33. Minimum temperature towards germination of cotton is

(a) 18°C
(b) 20°C
(c) 16°C
(d) 28°C

34. Minimum temperature towards growth of cotton lies

(a) 21-27°C
(b) 10-15°C
(c) 15-20°C
(d) 30-35°C

35. The recommended seed rate for *deshi* cotton is

(a) 6-10 kg/ha
(b) 20-25 kg/ha
(c) 10-18 kg/ha
(d) 25-30 kg/ha

36. The optimum population of hybrid cotton varieties is around

(a) 10,000 plants/ha
(b) 20,000 plants/ha
(c) 40,000 plants/ha
(d) 50,000 plants/ha

37. Which is the concept of critical stage of irrigation ?

(a) When irrigation is not given reduction of yield is maximum
(b) When water absorption rate is maximum
(c) When water loss by the plant is maximum
(d) None of these

38. Which is the most commonly followed method of planting cotton in India ?

(a) Direct seeding in rows
(b) Transplanting on raised seed beds
(c) Direct seeding in puddled beds
(d) None of these

39. The first inter-specific hybrid cotton developed in India in the year 1972 was

(a) Hybrid-4 (b) Savitri

(c) Varalaxmi (d) None of these

40. Parentage of first inter-specific hybrid cotton developed in India

(a) Female-hirsutum x male-barbadense

(b) Female-hirsutum x male-arboreum

(c) Female-barbadense x male-hirsutum

(d) None of these

41. India has the largest area in cotton cultivation in the world which is about ?

(a) 20m ha (b) 30m ha

(c) 8m ha (d) 60m ha

42. Productivity of cotton is highest in

(a) Rajasthan (b) Punjab

(c) U.P. (d) Haryana

43. The optimum pH of soil for cotton cultivation is lies between

(a) 9-12 (b) 5.5-8.5

(c) 3-5 (d) 9-11

44. The recommended dose of N, P_2O_5 and K_2O kg /ha for cotton are

(a) 80-40-40 (b) 110-30-70

(c) 40-30-40 (d) 120-40-40

45. The recommended dose of N, P_2O_5 and K_2O kg /ha for hybrid cotton are

(a) 120-150 - 60-70 - 40 (b) 120-40-40

(c) 30-20-30 (d) 80-40-40

46. The flower bud in cotton is known as

(a) Boll (b) Square

(c) Capsule (d) None of these

47. The fibre in cotton is elongation of ________ of seed coat

(a) Epidermal cell (b) Capsule

(c) Cuticle (d) None of these

48. The short fibre of cotton is known as

(a) Lint (b) Fuzz

(c) Thread (d) Cuticle

49. The long fibre of cotton is called

(a) Lint (b) Fuzz

(c) Thread (d) Cuticle

50. Varalaxmi is a inter specific variety of cotton evolved at

(a) Udaipur (b) Nagpur
(c) Jorhat (d) Bikaner

51. Origin of *Corchorus capsularis* (White jute) is believed to be in the

(a) Africa (b) India
(c) China (d) USA

52. "Boll shedding" in cotton is due to

(a) Frequent irrigation
(b) Inhibitory effect of ABA (Abscisic asid)
(c) Heavy dressing of urea
(d) All of these

53. Cotton belongs to the

(a) Liliaceae family (b) Malvaceae family
(c) Leguminaceae family (d) Compositae family

54. Delinting of cotton seed may be done with

(a) Sulphuric acid (H_2SO_4) (b) Nitric acid (HNO_3)
(c) Citric acid ($C_6 H_8O_7$) (d) Hydrochloric acid (HCl)

55. Gujarat -67 and American nectariless are parents of cotton hybrid_________.

(a) Hybrid-4 (b) Varalaxmi
(c) Hybrid-6 (d) Vikaram

56. What is the optimum temperatures towards for cultivation if jute ?

(a) 37°C (b) 30°C
(c) 34°C (d) 20°C

57. What is the observed best temperature for retting ?

(a) 24 °C (b) 20 °C
(c) 34 °C (d) 50 °C

58. Botanical name of white jute is

(a) *Corchorus olitouius* (b) *Corchorus capsularis*
(c) *Chochorus hirsuttum* (d) *Corchorus*

59. Botanical name of tossa jute is

(a) *Corchorus olitouius* (b) *Corchorus capsularis*
(c) *Chochorus hirsuttum* (d) All of these

60. Which jute is eaised on well drained uplands

(a) Tossa (*Corchorus olitorius*) (b) White (*Corchorus capsularis*)
(c) Both (a) and (b) (d) None of these

61. Which is a parenchymatus fibre ?

(a) Cotton (b) Jute
(c) Silk (d) All of these

62. What is the recommended dose of N, P_2O_5 and K_2O kg/ha for olitorius jute ?

(a) 60-40-40 (b) 120-40-40
(c) 30-20-30 (d) 100-50-40

63. What is the recommended dose of N, P_2O_5 and K_2O kg/ha for capsularis jute ?

(a) 100-50-40 (b) 120-40-40
(c) 80-40-20 (d) 60-40-40

64. Which soil types are unsuitable for cultivation of jute ?

(a) Deep alluvial soils receiving silt from annual floods
(b) Clay loam soils
(c) Sandy soils, heavy clay soils and soil with low pH
(d) None of these

65. Which of the following crop rotations ate not practiced with jute crop under irrigated conditions

(a) Jute-paddy-mustard, jute-paddy-pulses, jute-paddy
(b) Jute paddy-potato, jute-paddy-wheat
(c) Cowpea-jute-potato, jute-paddy-barseem
(d) None of these

66. In India the productivity of sunhemp is highest in the state

(a) Bihar (b) Odisha
(c) West Bengal (d) U.P.

67. In India the maximum area of sunhemp is in the state

(a) U.P. (b) Odisha
(c) West Bengal (d) Haryana

68. In India the maximum production of sunhemp is in the state

(a) West Bengal (b) Odisha
(c) U.P. (d) Rajasthan

69. The inflorescence in sunhemp is

(a) Raceme (b) Umbel
(c) Panicle (d) Cymose

70. Sunhemp belongs to family

(a) Cruciferae (b) Gramineae
(c) *Malvaceae* (d) Leguminoseae

71. Green manuring of sunhemp adds approximately ________ to the soils on decomposition

(a) 15-30 kg/ha (b) 40-60 kg/ha
(c) 70-90 kg/ha (d) 110-130 kg/ha

72. Sunhemp belongs to the family

(a) Leguminoseae (b) Gramineae
(c) Compositeae (d) Cucurbitaceae

73. The items of sunhemp is ________ in colour

(a) Green (b) Yellow
(c) Black (d) White

74. The optimum period of sowing for Capsularis types (white jute) is

(a) March - April (b) May - June
(c) August - September (d) December - January

75. The ideal stage of harvest of jute for fibre purpose is before

(a) One month old (b) Flowering
(c) Small pod stage (d) None of these

76. Jute belongs to the family ________

(a) Gramineae (b) Compositeae
(c) Cruciferae (d) Tiliaceae

77. What is the best observed pH for jute cultivation?

(a) 5.0-5.5 (b) 6.0-7.5
(c) 7.5-8.0 (d) 8.0-9.0

78. The optimum time of sowing for capsularis jute is

(a) March - April (b) January - February
(c) May - June (d) June - July

79. The recommended plant to plant spacing for jute is ________

(a) 8-10 cm (b) 15-20 cm
(c) 20-25 cm (d) 25-30 cm

80. Jute belongs to the family

(a) Cucurbitaceae (b) Papaveraceae
(c) Leguminaceae (d) Tiliaceae

81. The best observed soil pH for jute cultivation is

(a) 5.0-6.0 (b) 6.0-7.5
(c) 7.5-8.5 (d) 8.5-9.5

82. The optimum time of sowing for capsularis jute is

(a) March-April (b) April-May
(c) May-June (d) June-July

83. The optimum time of sowing for olitorius jute is

(a) March-April (b) April-May
(c) May-June (d) June-July

84. For fibre purposes, jute may be harvested at

(a) After pod formation (b) After flowering
(c) Before pod formation (d) Before flowering

85. Retting is a process by which________

(a) The fibers in the bark get loosened and separated from woody stalk
(b) Fibres are classified in to various retted groups
(c) The good quality fibre in separated
(d) None of these

86. The botanical name of jute is

(a) *Corchorus sliqua* (b) *Corhorus capsularis*
(c) *Corchorus hirsutum* (d) None of these

87. Nearly 70% of area under jute cultivation is

(a) *Corchorus olitorius* (b) *Corchrus capsularis*
(c) *Corchorus hirsutum* (d) *Corchorus sliqua*

88. The centre of origin of *Corchorus capsularis* is

(a) India and Burma (b) America
(c) Africa (d) China

89. The centre of origin of *Corchorus litorius* is

(a) India (b) America
(c) Africa (d) China

90. In India the productivity of jute is highest in the state

(a) Assam (b) America
(c) Bihar (d) China

91. In India the maximum area and production of jute is in the state

(a) Rajasthan (b) Haryana
(c) W. Bengal (d) U.P.

92. Sea Island cotton belong to

(a) *Gossypium lobatum* (b) *Gossypium herbaceum*
(c) *Gossypium hirsutum* (d) *Gossypium tomentosum*

93. **Which proportion of cotton area in India is rainfed?**

(a) 50% (b) 75%
(c) 20% (d) 60%

94. **In case of rainfed cotton cultivated in India, the minimum rainfall requirement is**

(a) 50 cm (b) 100 cm
(c) 70 cm (d) None of these

95. **Hampi, Khandwa-2, Lakshami, Mahalakshami, Lohit and Reba-B50 are the recommended improved varieties of**

(a) Sunhemp (b) Dhaincha
(c) Jute (d) Cotton

96. **Varalaxmi is a inter-specific variety of cotton evolved at**

(a) Nagpur (b) Patna
(c) Bikaner (d) Cuttack

97. **In case of cotton the flower are formed on**

(a) Monopodial branches (b) Sympodial branches
(c) All the branches (d) None of these

98. **In case of cotton the flower are formed on**

(a) Lower branches (b) Upper branches
(c) All of branches (d) None of these

99. **Which of the following fibre crop has the highest cultivated area in the world?**

(a) Cotton (b) Jute
(c) Sunhemp (d) Dhaincha

100. **Cotton is originated in**

(a) China (b) Europe
(c) South America (d) India

101. **"Boll" shedding in cotton is due to**

(a) Plant is over loaded with bolls
(b) Inhibitory effect of ABA (Abscisic acid)
(c) Heavy dressing of urea
(d) Frequent irrigation

102. **Cotton belongs to the _________ family**

(a) Gramineae (b) Malvaceae
(c) Crucifereae (d) Leguminaceae

103. Delinting of cotton seed may be done with________.

(a) Sulphuric acid (H_2SO_4)
(b) Nitric acid (HNO_3)
(c) Acetic acid (CH_3COOH)
(d) Hydrochloric acid (HCI)

104. Tossa jute is to be sown at inter row spacing of ______

(a) 35 cm
(b) 20 cm
(c) 10 cm
(d) 15 cm

105. For fibre purposes, jute, may be harvested at any time

(a) After pod formation
(b) After flowering
(c) Before pod formation
(d) Before flowering

106. Retting is a process by which

(a) The fibres in the bark get loosened and separated from woody stalk
(b) The poor fibre stalk is sorted out
(c) The good quality fibre in separated
(d) None of these

107. What is the observed best temperature for retting ?

(a) 18°C
(b) 25°C
(c) 34°C
(d) 40°C

108. The botanical name of sunhemp is

(a) *Corchorus sliqua*
(b) *Corchorus olitorius*
(c) *Corchorus hirsutum*
(d) *Crotolaria juncea*

109. The centre of origin of sunhemp is

(a) India and Burma
(b) USA
(c) S. Africa
(d) China

110. Which soil types are unsuitable for cultivation of sunhemp?

(a) Well drained alluvial soils with sandy loam to loamy texture
(b) Sandy loam soils
(c) Clay loam soils
(d) Sandy soils, heavy clay soils with low pH

111. Retting of sunhemp takes ________ days in December

(a) 4-8
(b) 10-12
(c) 12-15
(d) 21-25

112. Retting of sunhemp takes ________days in September

(a) 3-7
(b) 10-12
(c) 15-18
(d) 21-25

113. Ratting of sunhemp is a ________ process

(a) Chemical
(b) Physical
(c) Mechanical
(d) Microbiological

114. Which of the following crop rotations are practiced with jute crop under rainfed conditions?

(a) Jute-paddy-potato, jute-paddy-wheat
(b) Jute-wheat
(c) Cowpea-jute-potato,jute-paddy-barseem
(d) None of these

115. Retting of jute is a

(a) Chemical process
(b) Physical process
(c) Mechanical process
(d) Microbiological process

116. The inflorescence in jute is

(a) Siliqua
(b) Panicle
(c) Cymose
(d) Umbel

117. Jute belongs to ________

(a) Tiliaceae family
(b) Liliaceae family
(c) Malvaceae family
(d) Cruciferae family

118. Jute cultivation area in India is approximately

(a) 0.8 m ha
(b) 3.5 m ha
(c) 5.0 m ha
(d) 2.5 m ha

119. Bad opening of bolls in cotton is known as

(a) Mareck
(b) Tirak
(c) Dead boll
(d) Capsule

120. Which is favourable for tirak?

(a) Hot and dry winds at fruiting
(b) Rains at fruiting
(c) Cold winds at fruiting
(d) All of these

121. Tirak is associated with

(a) Light sandy soils with N deficiency
(b) Sandy loam soils with salinity
(c) Both (a) and (b)
(d) None of these

122. What is a tirak?

(a) Fungal disease
(b) Physiological disorder
(c) Bacterial disease
(d) Viral disease

123. The recommended seed rate for American cotton is ________

(a) 6-8 kg/ha
(b) 15-25 kg/ha
(c) 10-15 kg/ha
(d) 28-32 kg/ha

124. The chromosome number in *deshi* cotton is

(a) n=13 (b) n=23
(c) n=43 (d) n=27

125. The chromosome number in American cotton is

(a) n=15 (b) n=26
(c) n=42 (d) n=18

126. Cotton belongs to family

(a) Gramineae (b) Cucurbitaceae
(c) Malvaceae (d) Liliaceae

127. Sujatha, the first commercial extra long *Gossypium barbadense* cotton was released in the year

(a) 1969 (b) 1990
(c) 1979 (d) 1977

128. The first commercial Hybrid cotton H-4 was released in the year

(a) 1960 (b) 1980
(c) 1950 (d) 1970

129. Consider the following pairs of crops and varieties and select the pair which is correctly matched

(a) Barley : Kalyan sona (b) Cotton : Sujata
(c) Cowpea : Pusa phalguni (d) Green gram : Amber

130. Delinting of cotton seed may be done with

(a) Sulphuric acid (H_2SO_4) (b) Citric acid ($C_6 H_8O_7$)
(c) Nitric acid (HNO_3) (d) Hydrochloric acid(HCl)

131. The insect boll-worm is commonly found on

(a) Sunhemp (b) Jute
(c) Cotton (d) Castor

132. The recently developed terminator technology has been used in

(a) Jute (b) Cotton
(c) Tobacco (d) Sunhemp

133. Which is a variety of arboreum cotton?

(a) Shamli (b) Hampi
(c) Sujata (d) None of these

134. Which is a variety of barbadense cotton ?

(a) Shamli (b) Sujata
(c) Digvijaya (d) None of these

135. Which is a variety of herbaceum cotton?

(a) Sujata (b) Hampi
(c) Digvijaya (d) None of these

136. Which is a variety of hirsutum cotton?

(a) Shamli (b) Hampi
(c) Sujata (d) None of these

137. Critical stages for irrigation to cotton is

(a) Flowering stage (b) Boll formation stage
(c) Both (a) and (b) (d) Maturity stage

138. The oil content in the cotton seed lies between

(a) 5-10% (b) 15-25%
(c) 30-35% (d) 35-40%

139. Sea Island cotton belong to

(a) *Gossypium tomentosum* (b) *Gossypium herbaceum*
(c) *Gossypium hirsutum* (d) *Gossypium lebatum*

140. What is the optimum time of sowing for seed crop of sunhemp ?

(a) January (b) May
(c) April (d) August

141. Sunhemp is to be sown at inter row spacing of

(a) 30 cm (b) 25 cm
(c) 10 cm (d) 15 cm

142. For fibre purposes, sunhemp, may be harvested at

(a) Pod formation (b) After flowering
(c) Before pod formation (d) None of these

143. Retting is a process by which

(a) The fibres in the bark get loosened and separated from woody stalk
(b) The poor fibre stalk is sorted out
(c) Fibres are classified into various retted groups
(d) None of these

144. What is the observed best temperature of retting of sunnhemp ?

(a) 10-15°C (b) 21-27°C
(c) 15-27°C (d) 30-35°C

145. What is the recommended dose of N, P_2O_5 and K_2O kg/ha for sunhemp

(a) 20-50-30 (b) 120-40-40
(c) 80-40-20 (d) None of these

146. ________ is an early variety of sunhemp moderately resistant to shoot borer disease

(a) M-19 (b) Naveen
(c) Bellary (d) All of these

147. What is the recommended seed rate for line sowing of sunhemp ?

(a) 10-15 kg/ha (b) 5-8 kg/ha
(c) 40-50 kg/ha (d) 20-25 kg/ha

148. What is the recommended seed rate for sowing by broadcasting of sunhemp

(a) 5-8 kg/ha (b) 10-15 kg/ha
(c) 50-60 kg/ha (d) 30-40 kg/ha

149. The usual plant to plant spacing recommended for sunhemp is________

(a) 8-10 cm (b) 16-20 cm
(c) 12-18 cm (d) 22-26 cm

150. Which is an early variety of white jute suitable of double cropping ?

(a) Sonali (b) Subuj sona
(c) Naveen (d) None of these

151. Which is a variety of tossa jute?

(a) Sonali (b) Subuj sona
(c) Naveen (d) None of these

152. What is the recommended seed rate for *C. capsularis*?

(a) 8-10 kg/ha (b) 4-6 kg/ha
(c) 15-18 kg/ha (d) 19-24 kg/ha

153. What is the recommended seed rate for *Corchorus olitorius*?

(a) 10-12 kg/ha (b) 4-6 kg/ha
(c) 15-20 kg/ha (d) 25-30 kg/ha

154. The suitable temperatures for jute lies between

(a) 21-30°C (b) 24-37°C
(c) 20-35°C (d) 15-20°C

Fill in the Blanks

1. Cotton crop belongs to family ________
2. The appearance of flower bud in cotton is termed as ________
3. Jute Agricultural Research Institute is located at ________
4. ________ occupies first place in area under cotton production.
5. Oil content of cotton seed is from________ to ________%

6. Origin of sunhemp is
7. Jute belongs to family ________
8. ________________ is called as white jute
9. Fruiting branches in cotton are called as ________
10. Germination percentage of cotton is ________
11. Retting and steeping is followed in ________
12. ________________ is a fibre and oilseed crop.
13. Most common method of jute sowing is ________
14. Chromosome number of *deshi* and American cotton n=______
15. Fruiting branches in cotton are called as________

True/False

1. Cotton is often cross pollinated. **(T)**
2. *G. barbadence* species of cotton has longest fiber **(T)**
3. Jute fibre is blended with plastics. **(F)**
4. Recently developed terminator technology has been used in cotton. **(T)**
5. First marketing society on cotton at Gadag of Karnataka was started in 1917. **(T)**
6. Right stage of harvesting jute for higher yield and quality is late pod stage. **(F)**
7. Optimum temperature required for jute growth and retting process is 34°C. **(T)**
8. Ipomoea is common weed associated with sunhemp. **(T)**
9. Seed crop of jute spaced wider than the fiber crop. **(T)**
10. White jute requires lesser nutrient than tossa jute. **(F)**

Answers Fill in the Blanks

1. Malvaceae
2. Square
3. Barrackpore
4. India
5. 15% – 20%
6. India
7. Tiliaceae
8. *C. capsularis*
9. Sympodial
10. 60%
11. Jute
12. Linseed
13. Broadcasting
14. 13 & 26
15. Sympodial

ANSWERS

1.(a)	2.(c)	3.(a)	4.(d)	5.(b)	6.(d)	7.(a)
8.(c)	9.(a)	10.(a)	11.(b)	12.(d)	13.(a)	14.(b)
15.(a)	16.(a)	17.(b)	18.(a)	19.(b)	20.(a)	21.(c)
22.(a)	23.(b)	24.(c)	25.(a)	26.(a)	27.(b)	28.(c)
29.(c)	30.(a)	31.(a)	32.(d)	33.(c)	34.(a)	35.(c)
36.(d)	37.(a)	38.(a)	39.(c)	40.(a)	41.(c)	42.(b)
43.(b)	44.(a)	45.(a)	46.(b)	47.(a)	48.(b)	49.(a)
50.(b)	51.(a)	52.(b)	53.(b)	54.(a)	55.(a)	56.(c)
57.(c)	58.(b)	59.(a)	60.(a)	61.(b)	62.(a)	63.(a)
64.(c)	65.(a)	66.(a)	67.(a)	68.(b)	69.(b)	70.(d)
71.(b)	72.(a)	73.(b)	74.(a)	75.(b)	76.(d)	77.(b)
78.(a)	79.(a)	80.(d)	81.(b)	82.(a)	83.(b)	84.(d)
85.(a)	86.(b)	87.(b)	88.(a)	89.(c)	90.(a)	91.(c)
92.(c)	93.(b)	94.(a)	95.(a)	96.(a)	97.(b)	98.(b)
99.(a)	100.(d)	101.(a)	102.(b)	103.(a)	104.(c)	105.(d)
106.(a)	107.(d)	108.(d)	109.(a)	110.(a)	111.(c)	112.(a)
113.(d)	114.(b)	115.(d)	116.(c)	117.(a)	118.(a)	119.(b)
120.(a)	121.(c)	122.(b)	123.(b)	124.(a)	125.(b)	126.(c)
127.(a)	128.(d)	129.(b)	130.(a)	131.(c)	132.(b)	133.(a)
134.(c)	135.(c)	136.(b)	137.(c)	138.(b)	139.(c)	140.(d)
141.(c)	142.(a)	143.(a)	144.(b)	145.(a)	146.(a)	147.(d)
148.(c)	149.(a)	150.(a)	151.(c)	152.(a)	153.(b)	154.(b)

17

Forage Crop Production

1. ***Pennisetum pedicellatum* is a botanical name of**
   (a) Phulva grass (b) Sudan grass
   (c) Dinanath grass (d) Napier grass
2. **Which one is a multicut variety of sorghum?**
   (a) Pant Chari -3 (b) M.P. Chari
   (c) U.P. Chari-1 (d) Pusa Chari-1
3. **Which one is the dual purpose variety of sorghum?**
   (a) Pant Chari -5 (b) CSH-13R
   (c) CSV - 15 (d) All of these
4. **In hybrid napier grass crude protein content is**
   (a) 5.0 % (b) 18.5 %
   (c) 14.5 % (d) 10.2 %
5. **Napier grass is multiples through**
   (a) Seed (b) Stem
   (c) Root (d) Both (b) and (c)
6. **Most important cultivated species of oats in India is**
   (a) *Avena sativa* (b) *Avena byzantina*
   (c) *Avena abyssisica* (d) *Avena brevis*
7. **For fodder purpose variety of oat grown is**
   (a) Kent (b) Algerian
   (c) UPO 50 (d) All of these
8. **Seed rate of fodder oats sowing is**
   (a) 40 kg ha^{-1} (b) 60 kg ha^{-1}
   (c) 100 kg ha^{-1} (d) 150 kg ha^{-1}

9. **Fodder yield generally obtained from one hectare of area in a year of berseem is**
(a) 10-20 t ha^{-1} (b) 20-30 t ha^{-1}
(c) 50-60 ha^{-1} (d) 80-100 ha^{-1}

10. **Which is a diploid variety of berseem?**
(a) Miscavi (b) Khandaravi
(c) Chhindwara (d) All of these

11. **Which climatic condition is not suitable for harvest and hay making of forage?**
(a) Bright sunny hours (b) High relative humidity
(c) Both (a) and (b) (d) Very low temperature

12. **Which preservative used for silage making?**
(a) Sodium metabisulphide (b) Sulphur dioxide
(c) Molasses (d) All of these

13. **World's leading producer of milk**
(a) India (b) USA
(c) S. Africa (d) China

14. **Optimum sowing time of paragrass is**
(a) June - July (b) July - August
(c) August - September (d) May - June

15. **Optimum seed rate of napier grass is**
(a) 8000 - 10000 sets / ha (b) 15000- 18000 sets /ha
(c) 20000 - 24000 sets / ha (d) 12000 - 15000 sets /ha

16. **Optimum seed rate of sudan grass for seed production purposes is**
(a) 10-12 kg/ha (b) 16-20 kg/ha
(c) 6-8 kg/ha (d) 12-15 kg/ha

17. **Optimum sowing time of sudan grass in**
(a) Last week of March (b) First fornight of April
(c) Second (d) Last week of February

18. **Optimum sowing time of rhodun Green is**
(a) May-June (b) July - August
(c) June-July (d) October - November

19. **Optimum sowing time of guinea Green is**
(a) June-July (b) May-June
(c) August-September (d) October-November

20. Which parasite is associated with lucerne?

(a) *Cuscuta* (b) *Orabanche*
(c) *Striga* (d) None of these

21. *Cuscuta* (Doddar) in lucerne field can be central practices by

(a) Removing with host plant and burning
(b) Removing before seed set
(c) Spraying crude oil (d) All of these

22. Which is the variety of *Cyamopsis tetragonoloba* (cluster bean) is used both as grain and fodder ?

(a) Durgapur safed (b) Guara 80
(c) HFG 156 (d) All of these

23. Which is the right stage of harvesting fodder cluster bean?

(a) Vegetative stage (b) Flowering stage
(c) Early pod stage (d) All of these

24. Which of the following is the tetraploid berseem variety ?

(a) Pusa Giant (b) T 678
(c) Both (a) and (b) (d) Vardan

25. Which rhizobium species used for treating berseem seeds ?

(a) *trifolii* (b) *meliloti*
(c) *japonicum* (d) None of these

26. Optimum seed rate of berseem fodder is

(a) 10-15 kg ha^{-1} (b) 15-20 kg ha^{-1}
(c) 25-30 kg ha^{-1} (d) 50-60 kg ha^{-1}

27. In the following which is the lucerne variety ?

(a) Moopa (b) Rambler
(c) Sirsa No.8 (d) All of these

28. Optimum sowing time of lucerne crop is

(a) June (b) October
(c) November (d) August

29. Fodder can be obtained from oats in 3-4 cutting is

(a) 350 - 450 q/ha (b) 300-400 q/ha
(c) 500-550 q/ha (d) 600-700 q/ha

30. Fodder can be obtained from oats in cutting is

(a) 500-600 q/ha (b) 450-550 q/ha
(c) 350 - 450 q/ha (d) None of these

31. Optimum seed rate of oat for seed production is

(a) 80-90 kg/ ha (b) 100-130 kg/ha
(c) 70-80 kg/ha (d) 60-65 kg /ha

32. Optimum seed rate of lucerne sowing by Howard method is

(a) 10-12 kg/ha (b) 16-20 kg /ha
(c) 8-10 kg /ha (d) 12-15 kg /ha

33. Optimum seed rate of lucerne sowing by broadcast method

(a) 12 -15 kg /ha (b) 10-12 kg /ha
(c) 16-20 kg /ha (d) None of these

34. Optimum seed rate of lucerne for seed production is

(a) 6-8 kg /ha (b) 12-15 kg /ha
(c) 10-12 kg /ha (d) 8-10 kg /ha

35. Optimum sowing time of oats is

(a) Mid of October to last week of October
(b) Mid of August to last week of September
(c) First week of November to last week of December
(d) First week of September to last week of September

36. Optimum sowing time of lucerne for fodder purpose is

(a) Last week of September to last week of October
(b) Mid of August to first week of September
(c) Mid of November to last week of November
(d) None of these

37. Optimum seed rate of fodder maize is

(a) 50-60 kg/ha (b) 35-40 kg/ha
(c) 25-30 kg/ ha (d) 60-70 kg /ha

38. Optimum seed rate of small seeded sorghum is

(a) 30-45 kg/ha (b) 20-25 kg/ha
(c) 30-50 kg /ha (d) None of these

39. Optimum seed rate of bold seeded sorghum is

(a) 20-35 kg/ ha (b) 30-40 kg /ha
(c) 40-50 kg /ha (d) 50-60 kg/ha

40. Which Rhizobium species used for lucerne seed treatment ?

(a) *Trifolii* (b) *Meliloti*
(c) *Japonicum* (d) None of these

41. Napier grass *Pennisetum purpureum* cannot withstand

(a) Water-logging (b) Saline soils
(c) Frost (d) All of these

42. Hybrid napier can be inter-cropped with

(a) Cowpea (b) Berseem
(c) Lucerne (d) All of these

43. Which hybrid napier variety is suitable for inter-cropping situation?

(a) NB 21 (b) Pusa Giant
(c) Both (a) and (b) (d) None of these

44. Which of the following is a good indicator for good hay?

(a) Green colour leaves (b) Pleasant aroma
(c) Free from pathogens (d) All of these

45. Which is the optimum stage of harvest of forage crops for hay making?

(a) Vegetative stage (b) Flowering stage
(c) Grain formation stage (d) Maturity stage

46. Optimum seed rate of dinanath grass is

(a) 8-10 kg /ha (b) 15-18 kg /ha
(c) 10-12 kg /ha (d) 12-14 kg /ha

47. Optimum sowing time of napier grass is

(a) October - November (b) May - June
(c) June - July (d) August - September

48. Optimum sowing time of dinanath grass is

(a) June - July (b) October - November
(c) August - September (d) April - May

49. Fertilizer requirement of Rhodes green is

(a) 60 kg N + 40 kg P_2O_5 + 30 kg K_2O/ha
(b) 120 kg N + 60 kg P_2O_5 + 50 kg K_2O/ha
(c) 50 kg N + 20 kg P_2O_5 + 20 kg K_2O/ha
(d) 20 kg N + 40 kg P_2O_5 + 30 kg K_2O/ha

50. Fertilizer requirement of dinanath grass is

(a) 160 kg N + 20 kg P_2O_5 + 40 kg K_2O/ha
(b) 150 kg N + 40 kg P_2O_5 + 30 kg K_2O/ha
(c) 120 kg N + 20 kg P_2O_5 + 20 kg K_2O/ha
(d) 100-120 kg N + 40-50 kg P_2O_5 + 30-35 kg K_2O/ha

51. Grass tetany is caused by deficiency of ________ is forage crops

(a) Mg (b) Fe
(c) Ca (d) Sn

52. ***Molybdenosis*, a disease in cattle, is caused by an imbalance of Mo and Cu in the diet, when Mo content of the forage is greater than______**

(a) 5 ppm (b) 30 ppm
(c) 15 ppm (d) 10 ppm

53. **Fertilizer requirement of *deshi* sorghum is**

(a) 50-60 kg N + 30-40 kg P_2O_5 + 30 - 35 kg K_2O/ha
(b) 70-80 kg N + 30-40 kg P_2O_5 + 30 - 40 kg K_2O/ha
(c) 30-40 kg N + 20 kg P_2O_5 + 20 kg K_2O/ha
(d) None of these

54. **Fertilizer requirement of hybrid sorghum is**

(a) 40 kg N + 20 kg P_2O_5 + 20 kg K_2O/ha
(b) 50-60 kg N + 30-40 kg P_2O_5 + 30 - 35 kg K_2O/ha
(c) 100-120 kg N + 50-60 kg P_2O_5 + 40-50 kg K_2O/ha
(d) None of these

55. **Optimum seed rate of oat for fodder purposes is**

(a) 50-60 kg/ha (b) 60-70 kg/ha
(c) 70-80 kg/ha (d) 80-100 kg/ha

56. **Optimum seed rate of Rhodes grass by line sowing method is**

(a) 10-14 kg/ha (b) 8-10 kg/ha
(c) 12-16 kg/ha (d) 15-20 kg/ha

57. **Optimum sowing time of *Rabi sorghum* in north India is**

(a) Second week of September to second week of October
(b) First week of November to second week of November
(c) Mid of October to end of October
(d) None of these

Fill in the Blanks

1. Lucerne is a perennial _________
2. Maize in known as king of _________
3. Temporary roots found in maize are known as _________
4. Indian Grassland and Forage Research Institute is located at _________
5. Best cereal fodder crop for silage preparation is _________
6. Queen of the fodder crops is _________
7. Toxic substance present in fodder sorghum is _________
8. Best sowing month of fodder oats is _________
9. The centre of origin of sorghum is_______

10. The Directorate of Sorghum Research is situated at ________
11. Sorghum is considered a camel crop because of ________
12. Berseem seeds are treated with ________
13. The fodder which has maximum protein content (on dry wt. basis) is ________
14. World's leading producer of milk is ________
15. Process of green forage preserved in dry form without affecting the quality is known as ________

True/False

1. Soil sickness is caused by jowar. **(T)**
2. Sorghum is considered as camel crop because of deep root system. **(F)**
3. Subabul is a tree fodder. **(T)**
4. The nitrogen fixing fodder tree is Neem. **(F)**
5. In India, berseem was introduced from Egypt in 1904. **(T)**
6. Lucerne is a native of Peria. **(T)**
7. Berseem is considered as king of the fodder crops. **(T)**
8. *Pennisetum purpureum* (Napier grass) was introduced in India in the year 1912 from South Africa. **(T)**
9. Berseem seeds are treated with *R. japanicum.* **(F)**
10. Hay and silage is the best way to supply fodder during lean period. **(T)**

Answers Fill in the Blanks

1. Forage legume
2. Fodder
3. Seminal roots
4. Jhansi
5. Maize
6. Lucerne
7. HCN
8. October
9. India and Africa
10. Hyderabad
11. Resistance to drought
12. *R. trifoli*
13. Lucerne
14. India
15. Hay making

ANSWERS

1.(c)	2.(b)	3.(d)	4.(d)	5.(d)	6.(a)	7.(d)
8.(c)	9.(d)	10.(d)	11.(c)	12.(d)	13.(a)	14.(a)
15.(d)	16.(c)	17.(d)	18.(c)	19.(a)	20.(a)	21.(d)
22.(d)	23.(c)	24.(c)	25.(a)	26.(c)	27.(d)	28.(b)
29.(c)	30.(c)	31.(d)	32.(a)	33.(a)	34.(d)	35.(a)
36.(a)	37.(a)	38.(b)	39.(d)	40.(b)	41.(d)	42.(d)
43.(c)	44.(d)	45.(b)	46.(b)	47.(c)	48.(a)	49.(a)
50.(d)	51.(a)	52.(a)	53.(a)	54.(c)	55.(d)	56.(b)
57.(a)						

18

Plantation Crop Production

1. **The centre of origin of tea is**

   (a) Burma and China (b) USA
   (c) S. Africa (d) India

2. **In India the productivity of tea is highest in the state**

   (a) Punjab (b) Assam
   (c) W. Bengal (d) Rajasthan

3. **The tea is a ________ plant**

   (a) Annual (b) Evergreen
   (c) Deciduous (d) Biennial

4. **The botanical name of tea is**

   (a) *Camellia sinensis* (b) *Commelina communis*
   (c) *Coffea arabica* (d) *Calycoosia petiolata*

5. **The tea is made of**

   (a) Tender leaves (b) Unopened buds
   (c) Both (a) and (b) (d) Stem

6. **Natu special Prabhat Vishwanath and Takakw are important varieties in Andhra Pradesh of**

   (a) Hookah tobacco (b) Lanka tobacco
   (c) Natu tobacco (d) Cigar wrapper tobacco

7. **Rustica tobacco is also called**

   (a) Calcutti tobacco (b) Hookah tobacco
   (c) FCV tobacco (d) Bidi tobacco

8. **A-145 GT6, DK and GT8 are important varieties of Gujarat of**

   (a) Cigar tobacco (b) Bidi tobacco
   (c) FCV tobacco (d) Chewing tobacco

9. Meenakshi BR2, I-64, S-1, PV2, AD1, Vairan, Dhangan and Bhagylakshmi are important varieties of Tamil Nadu of

(a) FCV tobacco (b) Chewing tobacco
(c) Cigar tobacco (d) Bidi tobacco

10. Lanka special is an important variety of Andhra Pradesh of

(a) Natu tobacco (b) Hokkah tobacco
(c) Lanka tobacco (d) Cigar wrapper tobacco

11. NP 219, C2, C304 , Patiyali, ST-1 and SK 417 are important varieties of U.P. of

(a) Bidi tobacco (b) FCV tobacco
(c) Huuka tobacco (d) Chewing tobacco

12. Economic life of a tea bush in northern India is lies between

(a) 80-90 years (b) 50-60 years
(c) 40-50 years (d) 20-30 years

13. Economic life of tea bush in southern India is lies between

(a) 30-40 years (b) 50-60 years
(c) 40-50 years (d) 80 or more years

14. When plucking in tea is restricted to the terminal bud and two expanded succulent leaves, it is known as

(a) Fine plucking (b) Green plucking
(c) Coarse plucking (d) Bud plucking

15. When plucking of tea includes terminal bud and 3 expanded succulent leaves, it is known as

(a) Bud plucking (b) Medium plucking
(c) Coarse plucking (d) Succulent plucking

16. A special ________NPK soluble fertilizer mixture is used in tea nursery

(a) 1:1:1 (b) 3:2:2
(c) 4:3:1 (d) None of these

17. A________ NPK fertilizer mixture is used for first three years in tea

(a) 4:3:3 (b) 3:2:1
(c) 4:2:1 (d) 1:2:1

18. How many years old tea nursery seedlings are used for transplanting in the field ?

(a) Less than 1 (b) 1.5
(c) 2.0 (d) 3.5

19. What is the normal spacing for tea ?

(a) 120x120 cm (b) 120x150 cm

(c) 150x100 cm (d) 120x75 cm

20. What is the normal plant population / ha for tea ?

(a) 100000 (b) 2000

(c) 30000 (d) 10000

21. Heavy nitrogen application ________________ of smoking type tobacco

(a) Increases yield and decreases quality

(b) Increases quality and decreases yield

(c) Increases both quality and yield

(d) No effect

22. Heavy nitrogen application ________ of chewing, *bidi* type tobacco

(a) Increases yield and decreases quality

(b) Increases quality and decreases yield

(c) Increases both quality and yield

(d) No effect

23. Better quality of flue cured tobacco is obtained from

(a) Partially N starved plants (b) P sufficient plants

(c) Excess N supplied plants (d) Excess P supplied plant

24. Potash should never be supplied through ________ to tobacco

(a) Muriate of potash (b) Sulphate of potash

(c) Both (a) and (b) (d) None of these

25. Coffee belongs to family

(a) Malvaceae (b) Rubiaceae

(c) Poaceae (d) Euphorbiaceae

26. The centre of origin of *Coffee camphora* is

(a) Uganda (b) China

(c) USA (d) India

27. The centre of origin of *Coffee liberia* is

(a) India (b) Ethopia

(c) Liberia (d) China

28. The centre of origin of coffee arabica is

(a) Uganda (b) Ethopia

(c) Liberia (d) India

29. The coffee is a ________

(a) Annual plant (b) Biennial plant

(c) Perennial plant (d) None of these

30. **Nicotine content of the leaves of *Nicotiana tabacum* lies between________**
   (a) 2.0-5.0% (b) 1-2%
   (c) 3.5-8.0% (d) 9-12%

31. **Tea bushes are plucked in northern India from April to December at**
   (a) Monthly intervals (b) Fortnightly intervals
   (c) Weekly intervals (d) Interval of 5 days

32. **Rainfall at the time of maturity of tobacco is not desired as it causes________**
   (a) Washing of gums and resins from leaf
   (b) Lodging
   (c) Rotting of leaves
   (d) All of these

33. **Tobacco belongs to the family**
   (a) Malvaceae (b) Liliaceae
   (c) Solanaceae (d) Leguminoseae

34. **The optimum time for sowing nursery of tobacco is ________**
   (a) First fortnight of October (b) Second fortnight of October
   (c) First fortnight of August (d) None of these

35. **In tobacco topping of inflorescence along with few leaves is known as**
   (a) Topping high (b) Topping low
   (c) Topping full (d) None of these

36. **Branching in tobacco is mainly due to**
   (a) Topping (b) Desuckering
   (c) Detopping (d) Suckering

37. **Removal of suckers in tobacco is known as**
   (a) Topping (b) Desukering
   (c) Suckering (d) Detopping

38. **To obtain the benefit of topping in tobacco ________ is necessary**
   (a) Detopping (b) Desuckering
   (c) Suckering (d) Topping

39. **Desuckering in not required in ________**
   (a) Bidi type tobacco (b) Wrapper type tobacco
   (c) Ciggarette type tobacco (d) Hookah type tobacco

40. **Which tobacco is treated with salt water ?**
   (a) Wrapper (b) Chewing
   (c) Bidi (d) All of these

41. Which is not a method of curing ?

(a) Fire (b) Sun
(c) Chemical (d) Air

42. Sand drown in tobacco is mainly due to

(a) Zinc (b) Calcium
(c) Magnesium (d) Chlorine

43. Thickening of lower leaves in tobacco is due to

(a) Zinc (b) Calcium
(c) Magnesium (d) Boron

44. The process of removal of flower heads along with few upper leaves in tobacco is

(a) Desuckering (b) Topping
(c) Detopping (d) Suckering

45. In the tobacco topping results in improved

(a) Quality (b) Size
(c) Texture (d) All of these

46. Which is a Vilayati type tobacco?

(a) *Nicotiana tobacum* (b) *Nicotiana rustica*
(c) *Nicotiana benthamiana* (d) *Nicotiana glauca*

47. Which of the following is a stage of curing ?

(a) Yellowing (b) Fixing colour
(c) Drying (d) All of these

48. Which of the following is the deciding factor for quality, curing and yield ?

(a) Temperature (b) Sunlight
(c) Humidity (d) All of these

49. Which ions results in poor burning quality of leaves?

(a) Chloride (b) Calcium
(c) Phosphorus (d) Zinc

50. Which type of soil is suitable for cultivation of *bidi* tobacco?

(a) Medium loams (b) Sandy to silt loam
(c) Black soils (d) All of these

51. India maximum average under tobacco cultivation is in the state

(a) Rajasthan (b) Haryana
(c) Andhra Pradesh (d) U.P.

52. The production of tobacco in India is highest in the state

(a) Rajasthan (b) Haryana
(c) Andhra Pradesh (d) Uttar Pradesh

53. The inflorescence in tobacco is

(a) Raceme (b) Panicle
(c) Umbel (d) Spikelets

54. Optimum temperature for germination of tobacco is

(a) 15°C (b) 18°C
(c) 28°C (d) 24°C

55. The recommended seed rate for tobacco is

(a) 4-8 kg /ha (b) 2-3 kg /ha
(c) 4-6 kg /ha (d) 9-11 kg /ha

56. In southern India tea is harvested throughout the year, but harvesting intervals are longer

(a) during April - May (b) during December - March
(c) during June - August (d) during October - November

57. Which of the following process are involved in manufacture of black tea

(a) Harvested tea is withered and rolled or distorted by rolling in tea rolling machines to expose the leaf juice to air for oxidation
(b) Stopping fermentation by passing current of hot air after optimum chemical change is reached
(c) Grading by passing sieves
(d) All of these process

58. What is that special operation which is done during tea manufacture so that the end product is green tea ?

(a) Steaming the shoots soon after plucking to destroy leaf oxidase the fermentation enzyme
(b) Slowing done rolling process
(c) Alternate rolling and jerking
(d) None of these

59. The lowest leaf of tea is smallest and ______ ?

(a) True (b) Pseudo
(c) Broad (d) None of these

60. The serrated leaves of tea with no tips are called

(a) Gol pat (b) Janam
(c) Pseudo leaf (d) True leaf

61. The oval leaves of tea with tips not serrated are called

(a) Gol pat
(b) Janam
(c) Pseudo leaf
(d) Narrow leaf

62. The phylloxy of tea is known as ________

(a) 2-5 system
(b) 5-8 system
(c) 4-6 system
(d) 6-8 system

63. The coffee is made of

(a) Leaves
(b) Unopened buds
(c) Seeds
(d) None of these

64. ________ and ________ are principal economic species of coffee plant cultivated in India

(a) *Coffee arabica* and *Coffee camphora*
(b) *Coffee liberia* and *Coffee camphora*
(c) *Coffee arabica* and *Coffee liberia*
(d) None of these

65. Although coffee can be grown on soils of almost all physical types the ideal soil for its cultivation are

(a) Well drained, deep and friable loams
(b) Forest soils rich in organic matter
(c) Red and lateritic
(d) Black soil

66. The botanical name of *Robusta coffee* is

(a) *Coffee arabica*
(b) *Coffee camphora*
(c) *Coffee liberia*
(d) *Calycosia petiolata*

67. The botanical name of coffee is

(a) *Camellia sinenisis*
(b) *Commelina communis*
(c) *Coffea arabica*
(d) *Calycosia petiolata*

68. Tea is propagated by

(a) Seed and doves
(b) Root cuttings
(c) Leaf buds
(d) None of these

69. The normal spacing for *Arabica coffee* is

(a) 2×2 m
(b) 1×1 m
(c) 4×4 m
(d) 5×5 m

70. The normal spacing for Robusta coffee is

(a) 2×1 m
(b) 1×1 m
(c) 3×3 m
(d) 4×2 m

71. Directed spray of______ herbicide is recommended for weed control in coffee

(a) Glyphosate (b) 2,4-D
(c) Simazine (d) All of these

72. Which is the most important aspect for quality in coffee ?

(a) Picking (b) Roasting
(c) Nipping (d) None of these

73. Which seeds produces foxy coffee ?

(a) Under ripe (b) Over ripe
(c) Over roasted (d) None of these

74. The removal of green fruits after harvesting is known as

(a) Nipping (b) Dipping
(c) Stripping (d) Picking

75. Which type planting is adopted on horizontal slopes ?

(a) Vertical planting (b) Contour planting
(c) Line planting (d) None of these

76. In northern India harvesting of tea is suspended from December to March because in the above months are

(a) Treated with pesticides (b) In bearing
(c) Being trained (d) Dormant stage

77. What is the ideal pH for Tea cultivation ?

(a) 5.0 (b) 5.5
(c) 6.0 (d) 7.0

78. Which plant is planted to protect tea from high speed winds ?

(a) *Grewia robusta* (b) *Tectona grandis*
(c) *Dalbergia sissoo* (d) *Azadirachta indica*

79. To protect the soil from erosion in tea gardens ________ is done

(a) Contour terracing (b) Contour lining
(c) Bench terracing (d) Bench lining

80. In tea plucking is performed whole year

(a) NE India (b) South India
(c) Both (a) and (b) (d) None of these

81. Directed spray of ________ herbicide is recommended for weed control in tea

(a) Glyphosate (b) 2,4-D
(c) Simazine (d) None of these

82. Which is the most important insect causing about 75% loss in tea ?

(a) Tea mosquito (b) Red spider mite

(c) Termite (d) None of these

83. The centre of origin of tobacco is

(a) India (b) America

(c) S. Africa (d) China

84. In India the productivity of tobacco is highest in the state

(a) Rajasthan (b) Maharashtra

(c) Andhra Pradesh (d) Uttar Pradesh

85. Higher yield and quality of tobacco could be achieved by :

(a) Use of higher dose of phosphate (b) Adequate amount of Nitrogen

(c) Adequate irrigation (d) Liberal amount of K

86. The centre of origin of tobacco is

(a) India (b) America

(c) S. Africa (d) China

87. Transplanting of seedlings is commonly practiced in

(a) Tabacco (b) Wheat

(c) Barley (d) Oat

88. In tobacco the topping results in diversion of energy and nutrients to

(a) Leaves (b) Stem

(c) Fruits (d) None of these

89. The effect of topping in tobacco is more pronounced in

(a) Younger leaves (b) Middle leaves

(c) Older leaves (d) All of these

90. Topping of inflorescence in known as

(a) Topping high (b) Topping low

(c) Topping full (d) All of these

91. Flue cured tobacco is raised with low level

(a) N (b) P

(c) K (d) Zn

92. Which is not a stage of curing ?

(a) Yellowing (b) Fixing colour

(c) Hardening (d) Drying

93. Air Cured tobacco is

(a) Wrapper (b) Chewing

(c) Bidi (d) None of these

94. Fire cured tobacco is

(a) Wrpper (b) Chewing
(c) Bidi (d) None of these

95. Ideal pH value for coffee cultivation ________

(a) 4.5 (b) 5.5
(c) 6.0 (d) 7.5

96. Which is planted to protect coffee from high speed winds ?

(a) *Grewia robusta* (b) *Tectona grandis*
(c) *Delbergia sissoo* (d) *Magnifera indica*

97. How many years old coffee nursery seedlings are used for transplanting in the field ?

(a) Less than 1 (b) 1.5
(c) 5 (d) More than 2

98. Topping in tobacco means a process of ______

(a) Removal of all buds present in the axil of leaves
(b) Removal of leaves
(c) Removal of terminal bud
(d) None of these

99. The basic aim of topping and de-suckering in tobacco is

(a) To protect the plants against lodging
(b) To enlarge branching
(c) To divert energy and nutrients from flower head to leaves
(d) None of these

100. Botanical name of Deshi type tobacco is

(a) *Nicotiana tobacum* (b) *Nicotiana rustica*
(c) *Nicotiana glauca* (d) *Nicotiana benthamiana*

101. Which type of soil is suitable for binder (Cigar) tobacco cultivation ?

(a) Medium loams (b) Sandy silt soils
(c) Heavy soils (d) Black soils

102. Which type of soil is suitable for cigarette tobacco cultivation ?

(a) Medium loams (b) Light sandy soils
(c) Heavy soils (d) None of these

103. Which type of soil is suitable for hookah tobacco cultivation ?

(a) Medium loams (b) Sandy soils
(c) Black soils (d) Sandy to silt loams

104. The botanical name of arabica coffee is

(a) *Coffee arabica* (b) *Coffee camphora*

(c) *Coffee liberia* (d) *Calycosia petiolata*

105. Coffee is propagated by

(a) Seed (b) Cuttings

(c) Leaf buds (d) Tender leaves

106. Operations involved in manufacture of black coffee are

(a) Harvested coffee is withered and rolled or distorted by rolling in coffee- rolling machines to expose the leaf juice to air for oxidation (fermentation)

(b) Stopping fermentation by passing current of hot air after optimum chemical change is reached

(c) Grading by passing sieves

(d) All of these

107. Coffee prefers ________

(a) Hot and humid climate (b) Cool and dry climate

(c) Warm and dry climate (d) Cool and humid climate

108. Sonar Motihari DD 402, DD 403 and DD 404 are important varieties in West Bengal of

(a) Hookah tobacco (b) Bidi tobacco

(c) FCV tobacco (d) Chewing tobacco

109. Burley 21, KY 416, D1 and D2 are important varieties in Andhra Pradesh of

(a) FCV tobacco (b) Bidi tobacco

(c) Burley tobacco (d) Hookah tobacco

110. Nicotine based pesticides are

(a) Non persistent (b) Non systematic

(c) Contact insecticide (d) All of these

111. Tobacco oil can be safely used as

(a) For edible (b) For paint industry

(c) Both (a) and (b) (d) None of these

112. Nicotine sulphate and solanesol can be obtained from

(a) Tea (b) Coffee

(c) Tobacco (d) None of these

113. The harmful constituents in tobacco are

(a) Nicotine and tar

(b) Nitrosoamines and benzopyrenes

(c) Carbon monoxide and nitric oxides

(d) All of these

114. Chloride application ________ in tobacco.

(a) Decreases quality (b) Increases quality
(c) No effect no quality (d) Uncertain

115. Thick leaves and brostening of leaves in tobacco is due to

(a) Chlorine (b) Boron
(c) Zinc (d) All of these

116. Thin leaves and increased yield in tobacco is due to

(a) Zinc (b) Calcium
(c) Magnesium (d) Chlorine

117. By which process the harvested tobacco leaf is made ready for the market ?

(a) Curing (b) Loading in the barn
(c) Bulking (d) None of these

118. The recovery of oil from chewing tobacco is

(a) 30% (b) 35%
(c) 37% (d) 48%

119. The harmful constituents in tobacco can be minimized by application of

(a) Phosphatic fertilizers (b) Nitrogenous fertilizers
(c) Potassic fertilizers (d) None of these

120. Optimum seed rate of Motihari tobacco grown is West Bengal is

(a) 7 kg / ha (b) 8 kg /ha
(c) 6 kg / ha (d) 4 kg / ha

121. Motihari and Jati tobacco mainly grown is

(a) Andhra Pradesh (b) Tamil Nadu
(c) Bihar (d) West Bengal

122. HDBRG tobacco also known as

(a) Motihari (b) White barley
(c) Jati (d) Pikka

123. Harvesting method of FCV tobacco is

(a) Leaf wise (b) While plant
(c) Gugro and Lilvo (d) All of these

124. Harvesting method of bidi tobacco is

(a) Gugro and Lilvo (b) Leaf wise
(c) Whole plant (d) All of these

125. Harvesting method of chewing tobacco is

(a) Leaf wise (b) Gugro and lilvo
(c) Whole plant (d) All of these

126. Tea belongs to family

(a) Theacae (b) Malvaceae

(c) Rubiaceae (d) Euphorbiaceae

Fill in the Blanks

1. __________________ species of coffee is cultivated for parchment coffee.
2. Tobacco introduced by Portuguese to India during _________
3. Single fruit tobacco contains _________ seeds.
4. Hookah and Chewing tobacco research station is situated at ______
5. Most common curing method of bidi tobacco is _________
6. Vector of tobacco mosaic virus is _______
7. Longest producer of tobacco is _________
8. Common curing method for cigarette tobacco is_________
9. Coffee belongs to family_________
10. Scientific name of coffee is _________
11. In India the productivity of tea is highest in _________.
12. The centre of origin of tea is _________
13. Tea prefers _________ climate.
14. For the cultivation of tobacco ideal pH value is _________
15. Inflorescence in tobacco is _________
16. Scientific name of *Deshi* tobacco is _________
17. Tobacco belongs to family _________
18. The fruit of tobacco is known as_______
19. Scientific name of vilayati type of tobacco is _________
20. Topping in tobacco means a process of _______ .

True/False

1. Striga is commonly observed parasitic weed in tobacco. **(T)**
2. Suncuring is the common method for curing of cigarette tobacco. **(F)**
3. Ideal nitrogen to nicotine ratio of tobacco is 0.6-0.7. **(T)**
4. The coffee is made of tender leaves. **(F)**
5. Coffee is propagated by seed. **(T)**
6. Tea belongs to family Theaceae. **(T)**

7. Tea is propagated by leaf buds. **(F)**
8. Ideal pH for tea cultivation is 5.5. **(T)**
9. Red spider mite is the most important insect causing about 75% loss in tea. **(T)**
10. In India the productivity of tobacco is highest in the state of Uttar Pradesh. **(T)**

Answers Fill in the Blanks

1. *Coffee arabica*
2. 1905
3. 8000
4. Pusa
5. Sundrying
6. White fly
7. China
8. Flue curing
9. Rubiaceae
10. *Camellia sinensis*
11. Assam
12. Burma and China
13. Warm and humid
14. 5.5
15. Raceme
16. *Nicotiana tobacum*
17. Solanaceae
18. Siliqua
19. *Nicotiana rustica*
20. Removal of terminal bud

Answers

1.(a)	2.(b)	3.(b)	4.(a)	5.(c)	6.(c)	7.(a)
8.(d)	9.(b)	10.(c)	11.(c)	12.(c)	13.(d)	14.(a)
15.(a)	16.(a)	17.(a)	18.(b)	19.(d)	20.(d)	21.(a)
22.(c)	23.(a)	24.(a)	25.(b)	26.(a)	27.(c)	28.(b)
29.(b)	30.(c)	31.(c)	32.(a)	33.(c)	34.(c)	35.(b)
36.(a)	37.(b)	38.(b)	39.(b)	40.(b)	41.(c)	42.(c)
43.(d)	44.(b)	45.(d)	46.(b)	47.(d)	48.(c)	49.(a)
50.(a)	51.(d)	52.(c)	53.(a)	54.(c)	55.(b)	56.(b)
57.(d)	58.(a)	59.(b)	60.(a)	61.(b)	62.(a)	63.(c)
64.(a)	65.(c)	66.(b)	67.(a)	68.(a)	69.(a)	70.(c)
71.(a)	72.(a)	73.(b)	74.(c)	75.(b)	76.(d)	77.(a)
78.(a)	79.(c)	80.(b)	81.(a)	82.(b)	83.(b)	84.(d)
85.(d)	86.(b)	87.(a)	88.(a)	89.(a)	90.(a)	91.(a)
92.(c)	93.(a)	94.(b)	95.(c)	96.(a)	97.(b)	98.(c)
99.(c)	100.(a)	101.(c)	102.(b)	103.(d)	104.(a)	105.(a)
106.(d)	107.(a)	108.(a)	109.(c)	110.(d)	111.(c)	112.(c)
113.(d)	114.(a)	115.(a)	116.(b)	117.(a)	118.(c)	119.(c)
120.(c)	121.(d)	122.(b)	123.(a)	124.(d)	125.(d)	126.(a)

www.ingramcontent.com/pod-product-compliance
Ingram Content Group UK Ltd.
Pitfield, Milton Keynes, MK11 3LW, UK
UKHW020234290926
15470UKWH00002B/545

9 789351 309192